Physics Beyond Spin One

Small Things and Vast Effects

Thomas J. Buckholtz

T. J. Buckholtz & Associates
Portola Valley, California
USA

Physics Beyond Spin One
Small Things and Vast Effects

Edition 1

At the time this edition was published,

- The author could be reached via e-mail at Thomas.J.Buckholtz@gmail.com.
- The author had a website at http://thomasjbuckholtz.wordpress.com.

Published by
T. J. Buckholtz & Associates
Portola Valley, California USA

Printed by
CreateSpace
Charleston, South Carolina USA

LCCN: 2014900140

Table of Contents

Preface

Welcome to *Physics Beyond Spin One: Small Things and Vast Effects*.

In this book, I propose physics theory. Some of the theory falls beyond traditional physics research. I hope some of the concepts will prove useful.

You can use this book to gain non-traditional interpretations of nature. You can use concepts presented herein to think about fundamental physics and new applications of mathematics. Perhaps you will gain new vantage points for addressing traditional issues. Possibly you will see opportunities to enhance on-going or planned work. Perhaps you will find concepts for other research. Perhaps you will try to verify, refute, or extend work herein.

About the title

The following contributed to my choosing the title for this book.

- This book discusses an attempt at physics research. Hence, physics.
- This book proposes phenomena associated with elementary particles having spins greater than 1. Hence, beyond spin one.
- Hence, *Physics Beyond Spin One*.
- This book features an attempt to develop math correlating with elementary particles and their properties. Hence, small things.
- The book features an attempt to develop math correlating with forces that govern the rate of expansion of the universe. Hence, vast effects.
- Hence, *Physics Beyond Spin One: Small Things and Vast Effects.*

Acknowledgments

Various people provided nudges leading to work I published in *Physics Small and Vast: Basic Interactions.* Subsequently, Kamal Hanna suggested I think about leptoquarks. I found Wikipedia helpful regarding, for example, physics issues people consider unresolved.

About my hopes

I hope people will use this work. I hope people will benefit from this work. I hope people will tell me of extensions to this work, shortcomings in the work, and developments to which the work contributes.

- Thomas J. Buckholtz

Portola Valley, California USA
January 2014

Dedication

To Helen Buckholtz
And, in memory of Joel and Sylvia J. Buckholtz

About the author

Dr. Thomas J. Buckholtz is the author or a coauthor for articles, books, chapters, or reports regarding physics, applied physics, mathematics, computer science, applied computing, computer-based games, software licensing, innovation, systems-thinking tools, the information age, information proficiency, service science, governmental service to the public, and the role of chief information officers.

He played pivotal roles in the following endeavors.

- Create lines of business for a $1 billion (annual revenue) business unit.
- Save $100 million per year for a $6 billion company.
- Pioneer three information technologies.
- Establish three information-technology marketplace business practices.
- Develop useful, leading-edge business, engineering, and scientific software.
- Double a two-person firm's revenue, for each of two consecutive years.
- Preserve 7 kilometers of Pacific Ocean coastline.
- Create an international service program.
- Improve governmental service (from all levels of government) for the American public.
- Create a grassroots line-of-business for a United States political party's National Committee.

Tom served in the following capacities.

- Executive leading a $1 billion business unit
- Corporate officer and advisor for startups
- Chief information officer (CIO) for a $10 billion enterprise
- Co-CIO for the United States federal government's Executive Branch
- Program leader advocating innovation, enhancing teamwork, and providing information technology throughout a $6 billion company
- Commissioner, United States General Services Administration
- Mathematician; Scientist; Engineer
- Professorial Lecturer; University Extension Instructor
- Speaker; Workshop provider; Author
- Business advisor; Innovation consultant

Dr. Buckholtz's clients and employers have included large and small enterprises in aerospace, agricultural research, biotech, business services, computing, defense, education, energy utilities, government, healthcare, high technology, innovation, insurance, Internet, law enforcement, politics, research and development, telecommunications, and venture capital.

Tom served on elected or appointed boards or in other volunteer capacities for a residential cooperative, a swim club, and organizations in academia, innovation, and public policy. For a successful United States presidential campaign, he served as a donor, fundraiser, policy-research committee member, speaker, alternate delegate at the candidate's party's National Convention, speakers bureau leader, county cochairman, and county representative at regional and statewide meetings. He served as co-producer and co-host for 250 interview-format television programs discussing business, charitable, community, educational, governmental, and political endeavors.

His education includes the following.

- Earn a B.S. in mathematics from the California Institute of Technology.
- Earn a Ph.D. in physics from the University of California, Berkeley.
- Complete business administration programs at Stanford University and the University of Michigan.

Notice

Small Things and Vast Effects

Thomas J. Buckholtz

Thomas.J.Buckholtz@gmail.com

Math models may resolve about 10 particle-physics and astrophysics problems. The models use harmonic-oscillator math. The models correlate with Standard Model basic particles. The models seem to correlate with the following.

A family of zero-mass bosons includes photons, gravitons, and spin-3 and spin-4 particles. Effects of the family govern the rate of expansion of the universe. Dark matter and dark energy consist of up to two kinds of stuff. One kind features peers of baryonic matter. The other kind includes fermions with spins 3/2 and 7/2. C, P, and T violations exceed amounts correlating with models limited to spins that do not exceed 1. Reactions led to matter/antimatter imbalance. Gravitons and some spin-1 bosons correlate with neutrino oscillations. Some ratios correlating with particle masses feature integers. Basic fermions have 3 generations. Possibly-infinite zero-point vacuum energy need not be a concern.

Keywords

- Axion
- Baryonic matter
- Boson
- Charge
- Clumping
- Color charge
- Cosmic microwave background (CMB)
- CP violation
- CPT symmetry
- Dark energy
- Dark matter
- Density of the universe
- Elementary particles
- Fermion
- Fundamental forces
- General relativity
- Generations
- Gluon
- Graviton
- Higgs boson
- Lasing
- Leptoquark
- Magnetic moment
- Magnetic monopole
- Mass
- Masses of elementary particles
- Matter/antimatter imbalance
- Neutrino masses
- Photon
- Quantum gravity
- Quantum harmonic oscillator
- P violation
- Rate of expansion of the universe
- Space-like behavior
- Spin
- Standard Model
- Strong interaction
- Time-like behavior
- Unified electromagnetism and gravity
- Vector potential
- Weak interaction

Part 1 Introduction

Section 1 Context

Abs.1.1 Traditional mathematical models do not adequately correlate with physics observations.

We note aspects of physics for which mathematical models fall short

People develop mathematical models that correlate with physics observations. People use such models to talk about past observations and to predict future observations.

The next items list needs for which people say mathematical models fall short.

Unmet needs	(1.1)
Provide models people can use to ...	(1.2)
• List possible basic particles that have not been observed	(1.3)
• Describe quantum gravity	(1.4)
• Unify quantum gravity and quantum electromagnetism	(1.5)
• Explain dark matter	(1.6)
• Explain dark energy	(1.7)
• Explain changes in the rate of expansion of the universe	(1.8)
• Explain baryon asymmetry (matter/antimatter imbalance)	(1.9)
• Explain the sizes of some symmetry violations (P, CP, ...)	(1.10)
• Explain neutrino oscillations	(1.11)
• Predict neutrino masses	(1.12)
• Explain the number, 3, of generations of fermions	(1.13)
• Determine whether magnetic monopoles ever existed	(1.14)
• Address the zero-point energy of the vacuum	(1.15)
• Interrelate masses of basic particles other than charged leptons	(1.16)

Section 2 Core

Abs.2.1 Quantum isotropic harmonic oscillator methods (IOM) may adequately correlate with physics observations for which traditional models do not correlate.

We contrast IOM and the Standard Model

People use the Standard Model to discuss some physics observations. The next items list needs for which people say the Standard Model may suffice.

Needs met by the Standard Model	(2.1)
Provide models people can use to ...	(2.2)
• List basic particles that have been observed	(2.3)
• Describe and unify the electromagnetic, weak, and strong interactions	(2.4)

People base the Standard Model on Lagrangian mathematical techniques.

We develop and use quantum isotropic harmonic oscillator methods, math, and models (IOM).

The Standard Model features basic particles for which S (spin/ħ) does not exceed 1.

IOM correlate with possible particles for which S can be as much as 4. A subset of IOM correlates with all known basic particles.

We note possible usefulness of IOM

IOM seem to provide new insight regarding items following item (1.1) and items following item (2.1). Perhaps, when fully developed, IOM will provide adequate insight.

We note the scope of this paper

People might say that this paper discusses IOM adequately to provide useful insight regarding items following item (1.1) and items following item (2.1).

Section 3 Comments

Abs.3.1 Sections in this paper compile lists of inputs to and results from research this paper describes.

We discuss compendia this paper contains

The next items note places where this paper performs various functions. The place column indicates the section that performs the activity the function column notes. The list column shows letters that partly label some statements in this paper. This paper includes statements with labels of the form ύ.ϊ.ό for which ύ is 3 letters, ϊ is a section number, and ό is a number. For example, the 3 letters Abs denote abstract. Statements labeled in the form Abs.ϊ.ό summarize contents of sections. This paper lists references regarding numeric data.

Function	Place	List	(3.1)
Summarize results	Section 25	Abs	(3.2)
List guesses	Section 25	Gss	(3.3)
List possible opportunities for observational or experimental research	Section 25	SOR	(3.4)
List possible opportunities for theoretical research	Section 25	STR	(3.5)
List references	Section 26	Ref	(3.6)

We discuss structural elements this paper uses

Starting with Part 2, each section before Part 7 contains up to 4 elements - section abstract, context, core, and comments. Starting with Part 2, the introduction to each part contains up to 3 elements - context, core, and comments.

Part 2 Some math, some physics, and some numbers

Context

We discuss traditional mathematics and traditional physics

People describe, via traditional mathematics, math and solutions for isotropic harmonic oscillators.

People describe, via traditional applications of Lagrangian math and mixing angles, models people correlate with elementary particles.

We anticipate math and models correlating with aspects of elementary particles

We anticipate discussing underutilized math and solutions for isotropic harmonic oscillators. We anticipate offering math related to isotropic harmonic oscillators as an approach parallel to use of Lagrangian math to correlate with aspects of elementary particles.

Core

We preview sections in this part

One section discusses math for quantum isotropic harmonic oscillators. We find possibly underutilized solutions. We provide possibly new notation.

One section builds bridges between models based on quantum isotropic harmonics oscillators and people's observations regarding nature.

One section reviews some numbers people measure and defines some numbers we use.

Section 4 IOM (quantum isotropic harmonic oscillator methods, math, and models)

Abs.4.1 We introduce IOM (quantum isotropic harmonic oscillator methods, math, and models).

Context

We note traditional applications of harmonic-oscillator math

People use harmonic-oscillator math to model the motion of a mass attached to a spring. People state quantum versions of such math. People use quantum harmonic-oscillator math to model interactions between objects. Here, people may solve for approximate energy levels. Such applications do not pertain to modeling to describe basic properties (such as charge or mass) of basic particles (such as electrons or quarks). People use quantum harmonic-oscillator math to discuss the vector potential and to discuss lasing. As far as we know, people tend not to use negative quantum numbers for quantum harmonic-oscillator math.

We anticipate some uses of harmonic-oscillator math

We correlate quantum-harmonic oscillator math with internal (or invariant) properties (such as charge or mass) of basic particles. For non-zero-mass basic particles, such applications model mass (for bosons) and abilities to interact via bosons (for fermions). For zero-mass particles, such applications model concepts including and paralleling the vector potential.

Sometimes, we do not solve for an energy-like number. We assume that such a quantity exists or is not relevant. We look for symmetries.

Core

We note 4 types of traditional quantum-mechanics representations

People provide math to describe amplitudes for quantum states. The next items list 2 approaches.

A QM-type-C approach features terms expressed as functions (wave functions) of spatial coordinates (4.1)
- Some such wave functions satisfy partial differential equations

A QM-type-D approach uses terms not expressed by using spatial coordinates (4.2)
- Some such terms satisfy conditions related to raising operators and lowering operators

People can base use of each approach on either of 2 types of coordinates.

A CO-type-L approach features linear coordinates (4.3)
- For 3-dimensional quantum mechanics, people might denote coordinates by x, y, and z

A CO-type-S approach features radial and angular coordinates (4.4)
- For 3-dimensional quantum mechanics, people might denote the radial coordinate by r, with $r=(x^2+y^2+z^2)^{1/2}$

The next items show typical coordinates people use. These coordinates apply for 3 spatial dimensions.

(4.5)

Coordinate system	Coordinates	Symmetry point	Distance from symmetry point
CO-type-L	x, y, z	x=y=z=0	$r=(x^2+y^2+z^2)^{1/2}$
CO-type-S	r, θ, φ	r=0	r

$$-\infty<ύ<\infty\text{, for } ύ = x, y, \text{ or } z \quad (4.6)$$

$$0\leq r<\infty,\ 0\leq\theta\leq\pi,\text{ and } 0\leq\varphi\leq 2\pi \quad (4.7)$$

We note approaches we use

The next items show paired approaches we use.

QM-type-DL denotes an approach that combines QM-type-D and CO-type-L (4.8)
QM-type-CS denotes an approach that combines QM-type-C and CO-type-S (4.9)
QM-type-CL denotes an approach that combines QM-type-C and CO-type-L (4.10)

We denote 2 numbers of dimensions. Some people may choose to associate the subscript p with momentum-like or space-like. Some people may choose to associate the subscript e with energy-like or time-like.

D_p denotes an integer, with $D_p>0$ (4.11)
D_e denotes an integer, with $D_e>0$ (4.12)

The next items denote 2 sets. Each set consists of a sequence of consecutive integers.

$$SDp = \{ 1, 2, ..., D_p{-}1, D_p \} \quad (4.13)$$
$$SDe = \{ 1{-}D_e, 2{-}D_e, ..., -1, 0 \} \quad (4.14)$$

The next item denotes the union of the 2 sets.

$$SD = SDp \cup SDe \quad (4.15)$$

We use IOM for which the next items pertain. Here, for item (4.17), we use QM-type-DL. Here, ϵ denotes belongs to (or, is a member of). Here, n_χ denotes the quantum number for oscillator χ. Regarding item (4.18), the difference in signs has significance. The reverse choice of signs can also work.

$$D_p \text{ is odd and } D_e \text{ is odd} \quad (4.16)$$
$$0 = Œ = \Sigma_{\chi \epsilon SD} \pm_\chi (n_\chi + 1/2) \quad (4.17)$$
$$\pm_\chi = +1 \text{ for } \chi \le 0 \quad (4.18)$$
$$\pm_\chi = -1 \text{ for } 1 \le \chi$$

We show a traditional solution

The next items pertain to IOM for a traditional ground state with 3 spatial dimensions. Item (4.21) contributes +3/2 to item (4.17). Item (4.22) contributes −3/2 to item (4.17).

$$D_e = 1 \quad (4.19)$$
$$D_p = 3 \quad (4.20)$$
$$n_0 = 1 \quad (4.21)$$
$$n_1 = n_2 = n_3 = 0 \quad (4.22)$$

We explore aspects of a QM-type-CS approach and find non-traditional solutions

The next items show math for a QM-type-CS approach. People call item (4.24) the Laplacian operator for D dimensions. People call item (4.31) the potential.

$$\xi \Psi(r) = (\xi_0/2) (-\eta^2 \nabla^2 + \eta^{-2} r^2) \Psi(r) \quad (4.23)$$
$$\nabla^2 = r^{-(D-1)} (\partial/\partial r)(r^{D-1})(\partial/\partial r) - \Omega r^{-2} \quad (4.24)$$

ξ and $\xi_0/2$ denote numbers (4.25)
$\Psi(r)$ denotes a wave function (4.26)
r denotes a variable, with dimensions of length (4.27)
η denotes a length (4.28)
Ω denotes a number (4.29)
D denotes a non-negative integer (4.30)

$$V = (\xi_0/2)\, \eta^{-2}\, r^2 \quad (4.31)$$

The next item pertains.

For D=1, some solutions feature the following (4.32)

- $\Omega=0$
- The range $-\infty < r < \infty$ pertains
- ψ has the form of a Hermite polynomial (in variable r) multiplied by $\exp(-r^{-2} / (2\eta^2))$

Work below tends not to use solutions people associate with (4.32).
The next item describes solutions other than solutions people traditionally associate with (4.32).

$$\psi(r) \propto r^{\nu} \exp(-r^{-2} / (2\eta^2)) \quad (4.33)$$

The next items pertain for all solutions for which item (4.32) does not pertain.

$$\xi = (D+2\nu)(\xi_0/2) \quad (4.34)$$
$$\Omega = \nu(\nu+D-2) \quad (4.35)$$

The next items pertain to traditional solutions.

ν is non-negative (4.36)
ν is an integer (4.37)
Ω is non-negative (4.38)

Each of the next items points to non-traditional solutions.

ν can be negative (4.39)
ν can be other than an integer (4.40)

For D>2, item (4.39) is necessary (but not sufficient) for the next item to pertain.

Ω can be negative (4.41)

We focus on solutions that normalize

We limit our attention to solutions that can be normalized.
The next item shows behavior of the r-related normalization integrand near r=0.

$$\Psi^*\Psi\, r^{D-1} \sim r^{D-1+2\nu} \exp(-2r^2 2^{-1}\eta^{-2})) \sim r^{D-1+2\nu}, \text{ for } r\sim 0 \quad (4.42)$$

The next items pertain to some solutions that normalize.

$$-1 < D-1+2\nu \quad (4.43)$$
$$-D/2 < \nu \quad (4.44)$$
$$\psi \text{ normalizes if (but not only if) ... } -(D/2) < \nu \quad (4.45)$$

The next item provides a definition of the Dirac delta function. [Ref.4.1]

$$\delta(r) = \lim_{\varepsilon\to 0+} (1/(2(\pi\varepsilon)^{1/2})) \exp(-r^2/(4\varepsilon)) \quad (4.46)$$

We make the following association.

$$4\varepsilon = \eta^2 \quad (4.47)$$

We assume that use of items (4.46) and (4.47) correlates with extending the range of integration. The next item shows an extended range of integration. Perhaps, people should consider that $r_e=\infty$.

$$-r_e \leq r < \infty \quad (4.48)$$
$$r_e > 0$$

For r<0, we note the possibility that the angular dependence of Ψ changes. For example, $\cos(\theta)$ for r>0 might become $\cosh(\theta)$ for r<0. We anticipate the possibility of products of exponentials and trigonometric functions.

The next item supplements item (4.45).

ψ normalizes if (but not only if) ... $-(D/2) = \nu$ (4.49)

We coin these terms.

Inside denotes $-(D/2) < \nu$ (4.50)
Edge denotes $-(D/2) = \nu$ (4.51)

For each of inside or edge, η can have any real value other than 0. Two sets of mathematical solutions exist. One set corresponds to $\eta>0$. The other set corresponds to $\eta<0$.

For an edge case with -2ν an even integer, for each solution set, potentially 2 solutions exist. (For example, for $\Omega=0$, one potential solution has $\Psi(-r)=\Psi(r)$. Another potential solution has $\Psi(-r)=-\Psi(r)$.)

We call a linear combination (of potential solutions) that normalizes a type-1 solution. We call a linear combination that does not normalize a type-2 solution.

We base the following item on considerations related to item (4.48).

Gss.4.1 For an edge case with -2ν an even positive integer, 1 type-1 solution exists. (4.52)

For an edge case with -2ν an odd positive integer, 2 square roots of r^ν exist. Potentially, for each solution set, 4 solutions exist.

Gss.4.2 For an edge case with -2ν an odd positive integer, 3 orthogonal type-1 solutions exist. (4.53)

People apply the next items to traditional $D_p = 3$, r>0 math.

$$D_p = 3 \quad (4.54)$$
$$D = 3 \quad (4.55)$$
$$S = \nu\text{, for some non-negative integer } \nu \quad (4.56)$$
$$\Omega = S(S+1) \quad (4.57)$$
$$2S+1 \text{ angular solutions pertain} \quad (4.58)$$

The next items extend traditional $D_p = 3$ math. $D \neq D_p$ is allowed, as is $D = D_p$.

$$D_p = 3 \quad (4.59)$$

$-D_p \leq 2\nu$, with 2ν being an integer	(4.60)
$\Omega = \nu(\nu+D-2)$, for some non-negative integer D	(4.61)
$\lvert\Omega\rvert = S(S+1)$, for some S with 2S being a non-negative integer	(4.62)
2S+1 angular solutions pertain	(4.63)

We tabulate and symbolize solution sets and solutions

The next items summarize results. Regarding the numbers of solutions sets, the leftmost factor of 2 comes from the existence of 2 cases, namely $\eta>0$ and $\eta<0$.

Type	-2ν	Solution sets	Orthogonal type-1 solutions per set	(4.64)
inside	even and >0	2(2S+1)	1	(4.65)
inside	odd and >0	2(2S+1)	1	(4.66)
edge	even and >0	2(2S+1)	1	(4.67)
edge	odd and >0	2(2S+1)	3	(4.68)

We provide notation for solution sets and solutions

The next items provide notation for solution sets.

The symbol s± denotes a solution set	(4.69)
2s is an integer, with $-S \leq s \leq S$	(4.70)
± is + for $\eta>0$ and is − for $\eta<0$	(4.71)

The next items provide notation for type-1 solutions.

The symbol s±ï denotes a solution	(4.72)
s± denotes the solution set	(4.73)
ï is an integer, with 1 ≤ ï ≤ the number of type-1 solutions in the solution set	(4.74)

We show a non-traditional solution

We apply a QM-type-CS approach to the example that items including and following item (4.19) show. The next items show a non-traditional solution. This is an inside solution. Here, $S \neq \nu$.

$D_p = 3$	(4.75)
$\nu = -1$	(4.76)
$D = 3$	(4.77)
$\Omega = \nu(\nu+D-2) = -1(0) = 0$	(4.78)
$S = 0$	(4.79)
$\Omega = S(S+1) = 0$	(4.80)

We consider this non-traditional solution to be the ground state. The next item pertains.

$$\xi = (1/2)\xi_0 \qquad (4.81)$$

Along with the non-traditional solution, the next items pertain. These solutions correlate with solutions items including and following item (4.19) suggest. Traditionally, people state that the S=0 solution below corresponds to the ground state. For the S=0 solution below, $\xi = (3/2)\xi_0$.

$$\xi = (D+2\nu)\,(\xi_0/2) = (3/2 + S)\xi_0 \qquad (4.82)$$

S is a non-negative integer (4.83)

$$\Omega = S(S+1) \qquad (4.84)$$

We define open and closed, for pairs of harmonic oscillators

We use QM-type-DL. We consider 2 oscillators. One oscillator has index ó. The other oscillator has index ú. Here ó ≠ ú. Here, the state of the oscillator pair can be a sum of components. Each component consists of a product of a non-zero complex number and a basis amplitude we symbolize by $|\, n_{ó}, n_{ú} >$.

We assume exactly 1 of the following (4.85)

- ó≤0 and ú≤0
- 1≤ó and 1≤ú

We say that the ó-and-ú pair of oscillators is open if each of $n_{ó}$ and $n_{ú}$ is a single integer (4.86)

- Examples include
  - $n_{ó}\geq 0$, $n_{ú}\geq 0$
  - $n_{ó}=-2$, $n_{ú}=-1$
  - $n_{ó}=-1$, $n_{ú}=-1$

We say that the ó-and-ú pair of oscillators is closed if the state of the oscillator pair is a linear combination of at least 2 states, each having $n_{ó} + n_{ú} = -1$ (4.87)

- An example is
  - $(1/2)^{1/2}\,(\,|\, n_{ó}=0, n_{ú}=-1 > + \,|\, n_{ó}=-1, n_{ú}=0 >\,)$

We use a single * to denote the state of an oscillator that participates in a closed pair.
The concept of closed allows, for example, considering solutions for differing D_p to be equivalent.
For example, we can consider the following $D_e=1$, $D_p=5$ solution ...

n_0	n_1	n_2	n_3	n_4	n_5	
1	0	0	0	*	*	

(4.88)
(4.89)

... to be equivalent to the following $D_e=1$, $D_p=3$ solution.

n_0	n_1	n_2	n_3
1	0	0	0

(4.90)
(4.91)

For each of these 2 solutions, the 2-and-3 pair is open.

We narrow the types of oscillator pairs we consider

For the remainder of this paper, we restrict the definition of oscillator pair. The next item shows this restriction.

For the ó-and-ύ oscillator pair, (4.92)

- ó denotes an even (possibly non-positive) integer
- ύ = ó + 1

We define IOM(b_1;b_2,b_3;b_4)

The next items define the terms IOM(b_2,b_3) and IOM(b_1;b_2,b_3;b_4) for cases in which D_e is odd and D_p is odd. Here and below, the M in IOM can denote methods, math, or models.

IOM(b_2,b_3) denotes IOM restricted so that the following apply (4.93)

- $b_2 = -D_e + 1$
    - b_2 denotes the minimum χ for which n_χ belongs to an open pair
- $b_3 = D_p$
    - b_3 denotes the maximum χ for which n_χ belongs to an open pair

IOM(b_1;b_2,b_3;b_4) denotes IOM(b_2,b_3) that may be extended to include closed pairs of oscillators (4.94)

- Here, $b_1 \le b_2$ and $b_3 \le b_4$
- Here, $b_2 - b_1$ is a non-negative even integer
- Here, $b_4 - b_3$ is a non-negative even integer
- The number $b_2 - b_1$ denotes the number of extra oscillators χ for which $\chi < b_2$
- The number $b_4 - b_3$ denotes the number of extra oscillators χ for which $\chi > b_3$
- The extra oscillators come in closed pairs
- The next examples pertain
    - For IOM(b_2-2;b_2,b_3;b_3), the (b_2-2)-and-(b_2-1) pair pertains and is closed
    - For IOM(b_2;b_2,b_3;b_3+2), the (b_3+1)-and-(b_3+2) pair pertains and is closed

The next item pertains.

IOM(b_2;b_2,b_3;b_3) = IOM(b_2,b_3) (4.95)

Comments

We explore uncertainty related to solutions

We use a QM-type-CS approach. The next item follows from item (4.23).

$$\xi = (\xi_0/2)\,(\,\eta^2 <p_r^2> + \eta^{-2} <r^2>\,), \text{ in which} \quad (4.96)$$
$$<p_r^2> = <-\nabla^2> \text{ and}$$
$$<ύ> \text{ denotes the expected value of } ύ$$

The next item follows from item (4.34).

$$D + 2\nu = \eta^2\langle p_r^2\rangle + \eta^{-2}\langle r^2\rangle \quad (4.97)$$

$\eta^{-2}\langle r^2\rangle$ does not depend on η. Similarly, $\eta^2\langle p_r^2\rangle$ does not depend on η. The 2 terms contribute equally.

The next items pertain for $\xi_0 \neq 0$.

$\langle p_r^2\rangle \times \langle r^2\rangle$ does not vary with changes in η^2, for $\eta^2>0$ (4.98)
For edge solutions, people might consider that $\langle r^2\rangle = 0$ (4.99)

We suggest research

STR.4.1 Complete mathematics related to type-1 solutions sufficiently to describe wave functions for edge cases for $D_p = 3$ and $\nu = -3/2$. (Possibly, extend the work to pertain to edge cases for other D_p and ν.)

STR.4.2 To what extent might people derive benefit from IOM for any 1 or more than 1 of the following? D can be an integer < 1. D can be other than an integer. 2ν can be other than an integer. $\pm_\chi$ can be other than +1 or −1.

STR.4.3 To what extent would it be useful, for D=3, for people to consider a quantum number s such that $S=(s-1)/2$? (Here, $S(S+1)=(1/4)\cdot(s^2-1)$. Here, perhaps, s is any non-zero integer.)

STR.4.4 Explore IOM for cases in which D_e is even and D_p is even.

STR.4.5 How might people improve or extend the technique for cataloging quantum approaches?

We list references

Ref.4.1 Wolfram Alpha, computational knowledge engine, Wolfram Alpha LLC, http://mathworld.wolfram.com/DeltaFunction.html.

Section 5 IOM and physics

Abs.5.1 We focus on IOM for which $\Omega = \pm S(S+1)$, with S=spin/ħ, 2S being an integer, and $0 \le S \le 4$.

Context

We note properties people associate with traditional physics

As far as we know, people interpret all experiments and observations as being consistent with the next items.

Each basic particle has a spin/ħ we can denote by S (5.1)
For each basic particle, 2S is a non-negative integer (5.2)
For each known basic particle, S = 0, 1/2, or 1 (5.3)
People expect that, if gravitons exist, S=2 for gravitons (5.4)

We anticipate use of IOM

The next items characterize a range we anticipate to be relevant.

For each basic particle, 2S is a non-negative integer (5.5)

For each basic particle, $0 \le S \le 4$ (5.6)

Core

We start to bridge between physics and IOM

People use the next item to model quantum states of physics particles and systems.

$$\Omega = S(S+1) \quad (5.7)$$

Based on items (4.35), (5.1), (5.2), and (5.7), the next item defines a relevant number of dimensions.

$$D_{*p} = 3 \quad (5.8)$$

We use QM-type-CS. In essence, people use items (5.9) and (5.10) to generate item (5.11). People say that item (5.11) describes energy levels for a 3-dimensional quantum isotropic harmonic oscillator.

$$D = D_{*p} = 3 \quad (5.9)$$
$$S = \nu \ge 0 \quad (5.10)$$
$$\xi = (S + 3/2)\, \xi_0 \quad (5.11)$$

Apparently, people overlook the following solution. [items including and following item (4.75)] We characterize this solution as inside.

$$\nu = -1 \quad (5.12)$$
$$\Omega = 0 \text{ and } S = 0 \quad (5.13)$$
$$\xi = (1/2)\, \xi_0 \quad (5.14)$$

Something similar happens for QM-type-DL representations. The following shows an n'-times excited state for χ=2. Here, n' denotes a non-negative integer. Here, n'=0 denotes the ground state.

n_0	n_1	n_2	n_3	(5.15)
n'	−1	n'	0	(5.16)

We discuss correlations between IOM and physics

The next item provides context for work below.

Gss.5.1 For basic elementary particles, subsets of IOM(−8;−8,9;9) pertain. Parameters b_2 and b_3 in IOM(b_1;b_2,b_3;b_4) correlate with spin. ν=−1 correlates with basic bosons. ν=−3/2 correlates with basic fermion particles. ν=−1/2 correlates with fermion fields. (5.17)

The next item provides a condition we think correlates with a useful definition of universe.

IOM(−8;−8,9;9) pertains (5.18)

The next items provide conditions we think correlate with spin/$\hbar$ of a basic particle.

D_{*p} = 3 pertains (5.19)
We let S denote spin/ħ of a basic particle (5.20)

- S≥0
- 2S is an integer

Here, b_2 denotes the maximum value of b_2 for which IOM(b_1;b_2,b_3;b_4) describes the particle (5.21)
Here, b_3 denotes the minimum value of b_3 for which IOM(b_1;b_2,b_3;b_4) describes the particle (5.22)
For even 2S, (5.23)

- S = maximum($-b_2/2$, $(b_3-1)/2$)

For odd 2S, (5.24)

- S = maximum($-(b_2+1)/2$, $(b_3-2)/2$), with $b_2 \le -2$ and $b_3 \ge 3$

The next items provide conditions we think necessary for solutions that correlate with non-zero-mass basic particles. These items correlate with items above.

IOM(−8;−8,9;9) pertains (5.25)

- For non-zero-mass basic bosons, IOM(−8;−8,3;3) pertains
- For basic fermions, IOM(−8;−8,9;9) pertains

D_{*p} = 3 pertains (5.26)
The set Sν = (5.27)

- { −1} for boson fields and particles
- { −1/2} for fermion fields
- { −3/2} for fermion particles

2S is (5.28)

- an even integer for boson particles
- an odd integer for fermion particles

$0 \le S \le 4$ (5.29)
$\Omega = \pm S(S+1)$ (5.30)
$\Omega = \nu(\nu+D-2)$, for ν∊Sν and some integer D (5.31)
$D \ge 1$ (5.32)

The next items provide conditions we think necessary for solutions that correlate with zero-mass basic particles and their related fields. People characterize these particles as bosons. For these particles, perhaps people need not distinguish between particles and fields.

IOM(−2;−2,9;9) pertains (5.33)
D_{*p} = 3 pertains (5.34)
The set Sν = (5.35)

- { −1} for boson fields and particles

2S is an even integer (5.36)
$1 \le S \le 4$ (5.37)

Comments

We discuss the terms relative and invariant

People consider measurements of time, position, and momentum to be relative to observers. People consider some other quantities, such as rest mass, to be invariant with respect to observers.

We correlate some IOM aspects with other concepts

People may find that the next items provide insight regarding a mapping of aspects people may associate with IOM into aspects people may associate with energy-momentum space people correlate with physics of the e-family of basic particles. [Section 8]

Indices for axes in energy-momentum space (for the 1-axis aligned with kinematic momentum)	Indices χ for harmonic oscillators n_{χ} relevant to e-family IOM	(5.38)
0	0	(5.39)
1	1	(5.40)
2	2, 4, 6, and 8	(5.41)
3	3, 5, 7, and 9	(5.42)

We suggest research

STR.5.1 To what extent would people find beneficial defining and using a value for a D_{*e}?

Section 6 Some physics numbers

Abs.6.1 The mass of a tauon may equal a number computed from 4 physics constants.
Abs.6.2 We define a series of lengths, including the Planck length, based on 4 physics constants.
Abs.6.3 We note invariant properties of basic particles and of objects.

Context

We note importance of considering numbers

People want physics models to reflect known numbers and predict yet-to-be-measured numbers.

We anticipate computing some physics numbers

In this section, we compute numbers we use elsewhere in this paper.

Core

We relate the mass of a tauon to relative strengths of electromagnetism and gravity

The next item characterizes the relative strengths of photons and gravitons. Here, q_e denotes the charge of an electron, $1/(4\pi\varepsilon_0)$ denotes the Coulomb constant, G_N denotes the gravitational constant, and

m_e denotes the mass of an electron. This calculation pertains for electrons and positrons, but not particles having charges or masses other than those for electrons and positrons. We base numbers (here and below) on items following item (6.35). The uncertainty-range is approximate.

$$\{(q_e)^2/(4\pi\varepsilon_0)\} / \{G_N(m_e)^2\} \approx 4.1649(1)\times10^{42} \quad (6.1)$$

The next items define β and β'. Here, m_{tauon} denotes the mass of a tauon.

$$(4/3)(\beta^6)^2 = \{(q_e)^2/(4\pi\varepsilon_0)\} / \{G_N(m_e)^2\} \quad (6.2)$$
$$\beta' = m_{tauon} / m_e \quad (6.3)$$

The next items estimate β and β'. The uncertainty-range for β is approximate. For β, we estimate an uncertainty-range somewhat compatible with the larger relative uncertainty-range item (6.1) shows.

$$\beta \approx 3.477139(8)\times10^3 \quad (6.4)$$
$$\beta' \approx 3.47715(31)\times10^3 \quad (6.5)$$

The next item pertains.

Gss.6.1 $\beta' = \beta$. (6.6)

We note a series of lengths pertaining to numbers related to electrons and positrons

The next item notes a traditional physics length, the Planck length. We provide a symbol, λ_4, for that length. We add to the traditional statement 2 factors, each of value 1. The first such factor is m_e^0. The second such factor is 2^0.

$$\lambda_4 = G_N^{1/2}\, m_e^0\, \hbar^{1/2}\, c^{-3/2}\, 2^0 \quad (6.7)$$

The next item applies a traditional formula to a property (mass) associated with electrons. The formula represents the Schwarzschild radius. We denote the Schwarzschild radius by R_S. Traditionally, people apply the Schwarzschild-radius formula to black holes. Traditionally, people do not apply the formula to objects people claim have not enough mass to form black holes. We add to the traditional statement 1 factor, with value 1. That factor is $\hbar^0$.

$$\lambda_3 = R_S = G_N^1\, m_e^1\, \hbar^0\, c^{-2}\, 2^1 \quad (6.8)$$

The next item shows the ratio of the above 2 lengths.

$$Z = \lambda_4 / \lambda_3 = G_N^{-1/2}\, m_e^{-1}\, \hbar^{1/2}\, c^{1/2}\, 2^{-1} \approx 1.1945\times10^{22} \quad (6.9)$$

The next item defines a series of lengths.

$$\lambda_\$ = \lambda_4 \cdot Z^{(\$-4)} \quad (6.10)$$

The next items show factors and values (for electrons and positrons). These approximate lengths are the products of the factors indicated by the five columns having labels γ-for-$ύ^γ$ (for some ύ). G_N denotes the gravitational constant. m_e denotes the mass of an electron. ħ denotes Planck's constant. c denotes the speed of light. Times are computed via time=length/c. The time-centric column shows the log-base-

10 of times. The time since the big bang is ~$10^{17.6}$ seconds. The $ column values indicate possibly interesting correlations between items and e-family bosons we symbolize by $e%&. [Section 8] The $' column values indicate possibly interesting correlations between items and properties of objects. In effect, $'=$−1.

$	$\lambda_\$$	Length (m)	Log_{10} (time (sec))	Concept	γ for G_N^γ	γ for m_e^γ	γ for $\hbar^\gamma$	γ for c^γ	γ for 2^γ	$'	(6.11)
		3.3×10^{53}	+45		−1.5	−4	2.5	0.5	−4		(6.12)
		2.7×10^{31}	+23		−1	−3	2	0	−3		(6.13)
		2.3×10^{9}	+0.88		−0.5	−2	1.5	−0.5	−2		(6.14)
5	λ_5	1.9×10^{-13}	−21	spin/mass	0	−1	1	−1	−1	4	(6.15)
4	λ_4	1.6×10^{-35}	−43	Planck length	0.5	0	0.5	−1.5	0	3	(6.16)
3	λ_3	1.4×10^{-57}	−65	R_S	1	1	0	−2	1	2	(6.17)
2	λ_2	1.1×10^{-79}	−87		1.5	2	−0.5	−2.5	2	1	(6.18)
1	λ_1	9.5×10^{-102}	−109.5		2	3	−1	−3	3		(6.19)
		7.9×10^{-124}	−132		2.5	4	−1.5	−3.5	4		(6.20)

The next item pertains.

Gss.6.2 We attach significance to $\lambda_\$$ for which a particle property has an exponent γ=0. (6.21)

Possibly the formula for λ_5 pertains to other than electrons and positrons. For Z and W bosons, a spin/mass length may have significance. For pions, a spin/mass length may have significance.

A Z-boson spin/mass length is ~2×10^{-18} meters. People measure spatial dependence for interactions mediated by the weak interaction. For a separation of ~10^{-18} meters between 2 interacting particles, the weak interaction and the electromagnetic interaction have similar magnitudes. At a separation of ~3×10^{-17} meters, the weak interaction is less by approximately a factor of 10^4. [Ref.6.1] (6.22)

Calculating using a spin/ħ of 1/2, a pion spin/mass length would be a factor 139.6/0.511 or 273.2 smaller than that for electrons. (Spin/ħ=1/2 pertains to quarks. For a pion, spin/ħ=0.) That length is approximately 0.70×10^{-15} meters. An experimental charge radius for charged pions is $0.78^{+0.09}_{-0.10}\times10^{-15}$ meters. [Ref.6.2] (6.23)

G_N is not a particle property. We note item (6.2).

Gss.6.3 Regarding λ_5, people can consider q_e to be a particle property for which $|q_e|^0$ pertains. (6.24)

The next items pertain to electrons.

$'	Factor for which γ=0	Property of an electron	Name of property	(6.25)
4	q_e	q_e	Charge	(6.26)
3	m_e	m_e	Property-3	(6.27)

$'	Factor for which γ=0	Property of an electron	Name of property	(6.25)
2	$\hbar$	$(g_S)\hbar/2$	Magnetic moment	(6.28)
1	±1	±1	Property-1	(6.29)

The next item discusses property-3 for an object. For a non-zero-mass basic particle, property-3 denotes mass.

Property-3 denotes the energy/c^2 that varied observers would assign to an object, independent of the relative motions of the object and the observers (6.30)

- People might say that property-3 is the rest mass or rest energy

The next items define candidates for property-1 of an object.

Property-1 denotes the number of fermions ("fermion count") comprising an object (6.31)

Property-1 denotes the net handedness/chirality of fermions ("net handedness") comprising an object (6.32)

The next item pertains to electrons.

Gss.6.4 The particle properties spin/$\hbar$ (S=1/2), charge (q_e), mass (m_e). magnetic moment ($(g_S)\hbar/2$, with g_S=2), fermion count (1), and handedness/chirality (left) characterize an electron. (6.33)

The next item pertains to objects in general.

Gss.6.5 The properties spin/$\hbar$, charge, property-3, magnetic moment, fermion count, and net handedness characterize an object. (6.34)

Comments

We show numbers from experiments

The next items represent results of experiments. For items (6.47) and (6.48), g_S=2. [Ref.6.3, Ref.6.4, and Ref.6.5]

Symbol	Units	Number	Description	(6.35)
q_e	C	$-1.602176565(35)\times10^{-19}$	charge of an electron	(6.36)
ε_0	F m^{-1}	$8.854187817\times10^{-12}$	permittivity of free space	(6.37)
G_N	m^3 kg^{-1} s^{-2}	$6.67545(18)\times10^{-11}$	gravitational constant	(6.38)
m_e	kg	$9.10938291(40)\times10^{-31}$	mass of an electron	(6.39)
m_e	MeV/c^2	0.510998928(11)	mass of an electron	(6.40)
$\hbar$	J s	$1.054571726(47)\times10^{-34}$	Planck's constant	(6.41)
c	m s^{-1}	2.99792458×10^8	speed of light in a vacuum	(6.42)
α	(no units)	$7.2973525698(24)\times10^{-3}$	fine-structure constant	(6.43)
α^{-1}	(no units)	137.035999074(44)		(6.44)

Symbol	Units	Number	Description	(6.35)
m_{tauon}	MeV/c^2	$1.77682(16)\times10^3$	mass of a tauon	(6.45)
m_{muon}	MeV/c^2	105.6583715 ± 0.0000035	mass of a muon	(6.46)
a	(no units)	$(1159.65218076 \pm 0.00000027)\times10^{-6}$	electron magnetic moment anomaly $(g - g_S) / g_S$	(6.47)
a	(no units)	$(11659209\pm6)\times10^{-10}$	muon magnetic moment anomaly $(g - g_S) / g_S$	(6.48)

Values used below for ranges of quark masses come from [Ref.16.1].

We restate a formula for the mass of a tauon

The next item restates results from above.

$$m_{tauon} / m_e = \beta = \exp(\ (1/12)\ \log\{\ \{3/4\}\ \{(q_e)^2/(4\pi\varepsilon_0)\}\ /\ \{G_N(m_e)^2\}\ \}\ ) \qquad (6.49)$$

We note another number

The next items pertain.

$$Z' = (1/4\pi\varepsilon_0)^{-1/2}\,|q_e|^{-1}\,\hbar^{1/2}\,c^{1/2}\,2^{-1} \approx 5.8531 \qquad (6.50)$$

$$\alpha = (1/4)\cdot(Z')^{-2} \qquad (6.51)$$

We introduce notation we use regarding charge

The next item defines Q', a symbol related to charge. Here, Q denotes the charge of an object.

$$Q' = Q\ /\ |q_e| \qquad (6.52)$$

We suggest research

SOR.6.1 Verify (to a smaller than current experimental uncertainty-range) or refute $\beta' = \beta$.

We list references

Ref.6.1 Particle Data Group, Electroweak (web page), *The Particle Adventure*, Lawrence Berkeley National Laboratory, http://www.particleadventure.org/electroweak.html.
Ref.6.2 G. T. Adylov, et. al., A measurement of the electromagnetic size of the pion from direct elastic pion scattering data at 50 GeV/c, *Nuclear Physics B*, Volume 128, Issue 3, 3 October 1977, pages 461-505. (http://dx.doi.org/10.1016/0550-3213(77)90056-6)
Ref.6.3 T. Quinn et al, Improved Determination of G Using Two Methods, *Phys. Rev. Lett,* 111, 101102, 2013. (http://link.aps.org/doi/10.1103/PhysRevLett.111.101102)
Ref.6.4 J. Beringer et al. (Particle Data Group), *Phys. Rev. D86*, 010001 (2012). (http://pdg.lbl.gov/2012/reviews/rpp2012-rev-phys-constants.pdf)
Ref.6.5 J. Beringer et al. (Particle Data Group), *Phys. Rev. D86*, 010001 (2012). (http://pdg.lbl.gov/2012/tables/rpp2012-sum-leptons.pdf)

Part 3 Basic bosons and basic fermions

Context

We discuss traditional models for particles

People describe difficulties regarding correlating the Standard Model with nature. For example, people posit that gravitons mediate interactions people associate with gravity. People say the Standard Model does not provide for gravitons.

We anticipate a new model for particles

We anticipate correlating IOM with all known basic bosons, with gravitons, and with bosons yet to be discovered. We anticipate correlating IOM with all known basic fermions and with possible fermions yet to be discovered.

Core

We preview sections in this part

One section lists families of basic particles.

One section discusses a family of zero-mass basic particles including photons and gravitons.

One section discusses families of non-zero-mass basic particles related to and including the Z boson, W bosons, and Higgs boson.

One section discusses a family of particles related to gluons.

One section discusses families of basic particles related to and including leptons and quarks.

Section 7 Families of basic particles

Abs.7.1 A catalog of families of basic particles correlates with possible yet-to-be-discovered particles.

Context

We note that techniques for cataloging particles exist

Traditional catalogs of basic particles emphasize concepts such as boson or fermion, spin, and charge.

We note that we provide models that correlate with additional particles

We anticipate describing models correlating with types of possible particles traditionally not catalogued. For example, models correlate with possible fermions for which $S \neq 1/2$.

Core

We show a catalog of types of basic particles

The next items provide bases for cataloging basic particles.

For a basic particle, S denotes spin/ħ (7.1)

- 2S is a non-negative integer

For a basic particle, exactly 1 of the following pertains (7.2)

- 2S is even
- 2S is odd

For a basic particle, exactly 1 of the following pertains for a parameter Ω (7.3)

- $S>0$ and $\Omega = +S(S+1)$
- $S=0$ and $\Omega = 0$
- $S>0$ and $\Omega = -S(S+1)$

For a basic particle, m denotes mass and exactly 1 of the following pertains (7.4)

- $m = 0$
- $m \neq 0$

Items above correlate with possibilities for 12 families of basic particles. Here $12=2\times3\times2$.

The next items define notation for 7 families. Each known basic particle correlates with a family. We predict possible yet-to-be-discovered basic particles correlating with some of these families. We do not predict basic particles correlating with the 5 other possible families. The examples column lists some known particles. The examples column also lists the (hypothetical) graviton.

2S	Ω	Mass	Traditional theme	Symbol for the family	Examples of basic particles	(7.5)
Even	>0	0		e	Photon, graviton	(7.6)
Even	<0	0	Strong	s	Gluon	(7.7)
Even	>0	≠0	Weak	w	Z, W^-, W^+	(7.8)
Even	0	≠0	Higgs	h	Higgs	(7.9)
Even	<0	≠0		o		(7.10)
Odd	>0	≠0	Leptons	l	Electron, muon	(7.11)
Odd	<0	≠0	Quarks	q	Up, anti-up	(7.12)

Comments

We provide perspective regarding families of possible basic particles

The next items pertain to items following item (7.5).

Work in this paper correlates with possible yet-to-be-discovered basic particles in the following families (7.13)

- e-family
- o-family
- q-family

People may consider the e-family to be non-traditional (7.14)

- People may consider that the photon and graviton do not belong in 1 family

People may consider the o-family to be non-traditional (7.15)

- People may consider that no o-family particles have been discovered

Regarding the traditional theme column, ... (7.16)

- People might assign the term electromagnetic to a traditional interpretation of the e-family

Section 8 Electromagnetism, gravity, and so forth (the e-family)

Abs.8.1 The e-family includes photons, gravitons, and 2 other zero-mass basic bosons.
Abs.8.2 The e-family includes coherences of the family's 4 basic bosons.
Abs.8.3 Each e-family member has 2 modes (polarizations).
Abs.8.4 Each of the 4 e-family basic bosons mediates a force with spatial dependence R^{-2}.
Abs.8.5 E-family coherences provide forces with spatial dependences of R^{-4}, R^{-6}, and R^{-8}.
Abs.8.6 This paper may provide a way to avoid dealing with infinite photon ground-state energy.

Context

We note traditional difficulties regarding having a quantum theory of gravity

Traditionally, people have found difficulties in trying to develop quantum-mechanical theories of gravitation. People have had difficulty unifying electromagnetism and gravity. Here, unifying denotes developing useful quantum theory that encompasses both concepts. Such unity might, for example, feature a theoretical basis that points to each of the two interactions.

We anticipate unifying electromagnetism and gravity

We develop IOM that correlates with photons and gravitons belonging to a family that includes 4 basic zero-mass bosons. The model correlates with coherent phenomena, such as lasing and such as coherences between photons and gravitons.

Core

We model photons and the vector potential

People traditionally describe some aspects of a photon in terms of properties such as energy and momentum. Such properties vary based on the relative motions of observers.

People associate the vector potential with photons. People traditionally describe quantum mechanically the vector potential in terms of math correlating with excitations of 2 harmonic oscillators. This aspect has some invariance regarding observers. (To extent observers can choose differing sets of 2 directions orthogonal to the motion of a photon, this aspect in not completely invariant.)

We use QM-type-DL. The next item shows notation regarding modes (polarizations).

# denotes 0 excitation for an oscillator not participating in a mode (8.1)

The next items describe ground states for each of the 2 modes people attribute to photons. Here, we use IOM(0;0,3;3). Below, we explain notation denoting modes and particles.

n_0	n_1	n_2	n_3	Modes	(8.2)
0	−1	0	#	4e2	(8.3)
0	−1	#	0	4e3	(8.4)

These descriptions correlate with people's statements that a photon does not excite along its direction of motion. Here, n_1 correlates with the direction of motion. Applying an n_1-raising operator to either of these ground states results in a zero amplitude. Applying an n_1-raising operator to any excited state of such a ground state results in a zero amplitude.

The next items describe some excited states for n_2-polarization. People might say that the energy is proportional to $n_0 + 1/2$.

n_0	n_1	n_2	n_3	Mode	(8.5)
1	−1	1	#	4e2	(8.6)
2	−1	2	#	4e2	(8.7)
3	−1	3	#	4e2	(8.8)

We correlate, for zero-mass bosons, n_1 with a particle's direction of motion

The next item pertains.

For a zero-mass boson, people can consider that the $\chi=1$ oscillator correlates with the direction of momentum (or of motion) of the boson. (8.9)

We correlate n_1 with a boson's mass

The next item pertains.

Gss.8.1 For a basic boson, $n_1<0$ correlates with the boson's having no mass and $n_1\geq0$ correlates with the boson's having non-zero mass. (8.10)

We correlate, for zero-mass bosons, n_1 with spatial dependence of the related force

The next item describes forces intermediated by zero-mass bosons. For example, for photons, charge denotes the property for each object. For photons, $\acute{\upsilon}=2n_1=-2$. The condition $D_p>1$ excludes gluons.

Gss.8.2 For a basic particle with even 2S, with $D_p>1$, and with $n_1<0$, the force imparted between 2 non-overlapping objects scales as $R^{\acute{\upsilon}}$. Here, $\acute{\upsilon}=2n_1$. Here, R denotes the distance between the center of property of one object and the center of property of the other object. (8.11)

We develop a representation for gravitons

People guess that gravitons provide for the gravitational force. People say that gravitons have S=2, 2 polarization modes, and no mass. People say that gravity correlates with curvature in space time. As far as we know, no one has made a verified detection of a graviton.

Gss.8.3 $D_e=1$, $D_p=5$ solutions provide a model for gravitons. (8.12)

We use item (8.11). The next items model the ground state and first two excited states for n_4-polarized gravitons. Here, we use IOM(0;0,5;5).

n_0	n_1	n_2	n_3	n_4	n_5	Mode	(8.13)
1	−1	#	#	0	#	3e4	(8.14)
2	−1	#	#	1	#	3e4	(8.15)
3	−1	#	#	2	#	3e4	(8.16)

The next item models the ground state of n_5-polarized gravitons.

n_0	n_1	n_2	n_3	n_4	n_5	Mode	(8.17)
1	−1	#	#	#	0	3e5	(8.18)

We contrast models and data for photons and gravitons

The next items contrast models for photons and gravitons. Here, we use IOM(0;0,3;5) for photons. We use IOM(0;0,5;5) for gravitons.

For photons, the 4-and-5 oscillator pair is closed (8.19)
For gravitons, the 4-and-5 oscillator pair is open (8.20)

The next item interprets elements of the left side of item (6.2).

Gss.8.4 In the expression $(4/3)(\beta^6)^2 = \{(q_e)^2/(4\pi\varepsilon_0)\} / \{G_N(m_e)^2\}$, the leftmost exponent 2 represents the number of vertices in a Feynman diagram, β^6 represents the ratio of strengths per channel for electromagnetism and gravity (for an interaction between 2 electrons), 4 represents the number of channels for a photon, and 3 represents the number of channels for a graviton. (8.21)

We base the next item on items (8.19), (8.20), and (8.21).

Gss.8.5 A channel corresponds to a closed harmonic-oscillator pair. (8.22)

Beyond the 4-and-5 pair being closed for photons, it follows that, for each of photons and gravitons, 3 more closed χ-and-$(\chi+1)$ pairs pertain. Here, χ is even. For interactions that photons and gravitons carry, we consider IOM(−2;0,9;9).

We extend the series photons, gravitons, and so forth

We extend a series that begins with photons and gravitons. The next items show ground states for even-polarized non-zero-mass basic particles. We use IOM(−2;0,9;9). Thereby, we allow for channels.

n_{-2}	n_{-1}	n_0	n_1	n_2	n_3	n_4	n_5	n_6	n_7	n_8	n_9	Modes	(8.23)
*	*	0	−1	0	#	*	*	*	*	*	*	4e2	(8.24)
*	*	1	−1	#	#	0	#	*	*	*	*	3e4	(8.25)
*	*	2	−1	#	#	#	#	0	#	*	*	2e6	(8.26)
*	*	3	−1	#	#	#	#	#	#	0	#	1e8	(8.27)

The mode item (8.26) models correlates with 2 channels. The mode item (8.27) models correlates with 1 channel.

We discuss an issue regarding $D_p>3$

People perceive that local phenomena occur in a space time with 3 spatial dimensions. The next item discusses how modes for $D_p>3$ correlate with $D_p=D_{*p}=3$ physics. [items following item (5.38)]

For $\chi>3$, with χ denoting the number of an oscillator, (8.28)

- Modes having even χ exert influence in the same spatial direction as modes having $\chi=2$
- Modes having odd χ exert influence in the same spatial direction as modes having $\chi=3$

We provide notation for e-family members

We denote bosons closely related to the above series by \$e%&. Here, $D_p\geq3$. We use IOM(−2;0,9;9).

$\$ = 5 - (D_p-1)/2$ (8.29)

The symbol e denotes related to electromagnetism (8.30)

The list % contains even integers (8.31)

- Values for those integers can be 2, 4, 6, or 8
- Integers appear in ascending order
- The list contains no less than 1 element and no more than 4 elements
- No integer appears more than once
- Each integer χ that appears corresponds to an open χ-and-(χ+1) oscillator pair

The symbol & denotes boson (8.32)

- Similar notation without & (and possibly with only odd values for χ) describes modes

We list some particles, coherences, modes, and force spatial-dependences

The next items show some e-family ground states. We call these the photon-graviton series bosons.

n_{-2}	n_{-1}	n_0	n_1	n_2	n_3	n_4	n_5	n_6	n_7	n_8	n_9	Particles	(8.33)
*	*	0	−1	0	0	*	*	*	*	*	*	4e2&	(8.34)
*	*	1	−1	#	#	0	0	*	*	*	*	3e4&	(8.35)
*	*	2	−1	#	#	#	#	0	0	*	*	2e6&	(8.36)
*	*	3	−1	#	#	#	#	#	#	0	0	1e8&	(8.37)

The next items show ground states for odd-polarization modes of photon-graviton series bosons.

n_{-2}	n_{-1}	n_0	n_1	n_2	n_3	n_4	n_5	n_6	n_7	n_8	n_9	Modes	(8.38)
*	*	0	−1	#	0	*	*	*	*	*	*	4e3	(8.39)
*	*	1	−1	#	#	#	0	*	*	*	*	3e5	(8.40)
*	*	2	−1	#	#	#	#	#	0	*	*	2e7	(8.41)
*	*	3	−1	#	#	#	#	#	#	#	0	1e9	(8.42)

The next items show some e-family ground states. We call these the maximal-% e-family members. For other than 4e2%, the e-family members shown are coherences.

n_{-2}	n_{-1}	n_0	n_1	n_2	n_3	n_4	n_5	n_6	n_7	n_8	n_9	Particles or coherences	(8.43)
*	*	0	−1	0	0	*	*	*	*	*	*	4e2&	(8.44)
*	*	0	−2	0	0	0	0	*	*	*	*	3e24&	(8.45)
*	*	0	−3	0	0	0	0	0	0	*	*	2e246&	(8.46)
*	*	0	−4	0	0	0	0	0	0	0	0	1e2468&	(8.47)

The next items show ground states for even-polarization modes of maximal-% e-family members. We use the term coherence to describe modes that encompass more than 1 basic e-family boson. Items (8.50), (8.51), and (8.52) exemplify this concept of coherence.

n_{-2}	n_{-1}	n_0	n_1	n_2	n_3	n_4	n_5	n_6	n_7	n_8	n_9	Modes	(8.48)
*	*	0	−1	0	#	*	*	*	*	*	*	4e2	(8.49)
*	*	0	−2	0	#	0	#	*	*	*	*	3e24	(8.50)
*	*	0	−3	0	#	0	#	0	#	*	*	2e246	(8.51)
*	*	0	−4	0	#	0	#	0	#	0	#	1e2468	(8.52)

The next items show first excited states for maximal-% even-polarization modes.

n_{-2}	n_{-1}	n_0	n_1	n_2	n_3	n_4	n_5	n_6	n_7	n_8	n_9	Modes	(8.53)
*	*	1	−1	1	#	*	*	*	*	*	*	4e2	(8.54)
*	*	2	−2	1	#	1	#	*	*	*	*	3e24	(8.55)
*	*	3	−3	1	#	1	#	1	#	*	*	2e246	(8.56)
*	*	4	−4	1	#	1	#	1	#	1	#	1e2468	(8.57)

The next items show spatial dependences for forces correlated with some e-family members.

Particles or coherences	Spatial dependence of force	(8.58)
4e2&	R^{-2}	(8.59)
3e4&	R^{-2}	(8.60)
2e6&	R^{-2}	(8.61)
1e8&	R^{-2}	(8.62)
3e24&	R^{-4}	(8.63)
2e246&	R^{-6}	(8.64)
1e2468&	R^{-8}	(8.65)

We discuss object-properties for which e-family bosons couple basic fermions and other objects

The next items use the series items following item (6.25) present.

$e%& forces for which % contains a 2 couple to charge (8.66)
$e%& forces for which % contains a 4 couple to property-3 (8.67)
Gss.8.6 $e%& forces for which % contains a 6 couple to magnetic moment. (8.68)
Gss.8.7 $e%& forces for which % contains an 8 couple to property-1. (8.69)

Comments

We discuss coherences

When % contains more than 1 element, an excitation provides for coherence between the excitations of the oscillators corresponding to the various elements. Coherence need not imply identical observed magnitudes and directions of momenta for e-family basic particles that correspond to the various elements in a coherence. In Section 18, we discuss coherence for 2 photons produced via electron-positron annihilation. In Section 13, we discuss the possibility that 3e24& (or other coherences) correlates with phenomena people correlate with the concept of axions.

Possibly, the next items show raising operators. Here, n denotes the quantum number shared by the even-oscillators in open pairs.

Modes	Raising operators	
		(8.70)
4e2	$a^+ \mid n > = (1+n)^{1/2} \mid n+1 >$	(8.71)
3e24	$a^+ \mid n > = (1+n)^{1} \mid n+1 >$	(8.72)
2e246	$a^+ \mid n > = (1+n)^{3/2} \mid n+1 >$	(8.73)
1e2468	$a^+ \mid n > = (1+n)^{2} \mid n+1 >$	(8.74)

We discuss QM-type-DL representations and spin/ħ

People say that S for gravitons should be 2. The next item generalizes.

Gss.8.8 For e-family basic bosons, 2S = the maximum χ for which the χ-and-(χ+1) pair is open. (8.75)

We discuss spatial dependence of e-family forces

People correlate R^{-2} force behavior for 4e2& with the notion that, in 3-dimensional space, the areas of the surfaces of spheres increase in proportion the square of the radii of the spheres. Items including and following item (8.58) extend such thoughts.

We note the topic of the extent to which e-family bosons provide attraction and repulsion

Forces in the e-family can provide attraction or repulsion. In Section 12, we note data that may correlate with 2 maximal-% e-family bosons (other than 4e2&) providing repulsion and 1 maximal-% e-family boson (other than 4e2&) providing attraction.

We note we may have resolved the matter of zero-point energy

The next items present the possibility people can consider that this work resolves the matter of photon zero-point energy.

In traditional physics, the sum over photon states of ground-state energy is unbounded (8.76)

IOM feature Œ=0 for bosons (8.77)

We suggest research

SOR.8.1 Detect instances or effects of, or rule out (to some confidence level), 3e24& coherence between photons and gravitons.

SOR.8.2 Measure or infer signs and magnitudes for forces mediated by e-family members other than 4e2& and 3e4&.

STR.8.1 To what extent might people benefit by considering that, for IOM(−2;0,9;9), the 0-and-1 oscillator pair correlates with a fold in an otherwise flat energy-momentum space (or space time), that oscillators −2 through 0 correlate with 2 or 3 flat energy-like (or time-like) dimensions, and that oscillators 1 through 9 correlate with 8 or 9 flat momentum-like (or space-like) dimensions?

STR.8.2 To what extent might people find it appropriate to associate $(t')^0$ behavior with \$e%& interactions? (Here, t' denotes time.)

STR.8.3 Harmonize models and observations or experiments regarding S for e-family members.

Section 9 Non-zero-mass basic bosons (the w-, h-, and o-families)

Abs.9.1 Non-zero-mass basic bosons include the w-family (Z, W^-, and W^+), the Higgs boson, and o-family bosons (for which $\Omega<0$).

Context

We note some known and conjectured non-zero-mass basic bosons

People discuss non-zero-mass basic bosons. The Standard Model correlates with the Z, W^-, W^+, and Higgs bosons. People discuss a concept of leptoquarks. People perform experiments to try to detect leptoquarks.

We anticipate that IOM correlates with yet-to-be-detected non-zero-mass basic bosons

We discuss the o-family.

Core

We discuss IOM correlating with non-zero-mass basic bosons

We use QM-type-CS. Based on items (4.35) and (5.17), the next item pertains for non-zero-mass basic bosons.

$$\text{D for fields} = 3 - \Omega \qquad (9.1)$$

Based on known numbers of particles in the w- and h-families, the next item pertains.

$$\text{Number of particles} = 2S+1 \qquad (9.2)$$

We list non-zero-mass basic bosons

We correlate particles with IOM solutions for which ν=−1. The next items pertain. Rows for which the number of particles column shows a blank fall outside assumptions we make.

S	Ω	D	Traditional particles	Possible particles	Number of particles	(9.3)
2	6	−3				(9.4)
1	2	1	Z, W⁻, W⁺		3	(9.5)
0	0	3	Higgs		1	(9.6)
1	−2	5		4o%	3	(9.7)
2	−6	9		3o%	5	(9.8)
3	−12	15		2o%	7	(9.9)
4	−20	23		1o%	9	(9.10)
...	−S(S+1)					(9.11)

We denote non-zero-mass basic bosons by $ï%. We use IOM(−8;−8,3;3).

$ = 5 − S (9.12)

These symbols pertain (9.13)

- ï=w for the w-family
- ï=h for the h-family
- ï=o for the o-family

These numbers pertain (9.14)

- $=4 for the w-family
- $=5 for the h-family
- 4≥$≥1 for the o-family

% denotes an integer χ (9.15)

- The integer χ correlates with the harmonic oscillator with quantum number n_χ
- For the w-family, 1≤χ≤3
- For the h-family, χ=1 (or χ=0, which provides equivalent notation for the same particle - the Higgs boson)
- For the o-family, −8≤χ≤0

The next items note symbols.

Particles	Symbols	(9.16)
Z	4w1	(9.17)
W⁻	4w2	(9.18)
W⁺	4w3	(9.19)
Higgs	5h(0) and 5h1	(9.20)

The next items provide a QM-type-DL model for w-, h-, and o-family ground states.

n_{-8}	n_{-7}	n_{-6}	n_{-5}	n_{-4}	n_{-3}	n_{-2}	n_{-1}	n_0	n_1	n_2	n_3	n_4	n_5	n_6	n_7	n_8	n_9	Modes	(9.21)
*	*	*	*	*	*	*	*	1	0	#	#							4w1	(9.22)
*	*	*	*	*	*	*	*	1	#	0	#							4w2	(9.23)

n_{-8}	n_{-7}	n_{-6}	n_{-5}	n_{-4}	n_{-3}	n_{-2}	n_{-1}	n_0	n_1	n_2	n_3	n_4	n_5	n_6	n_7	n_8	n_9	Modes	(9.21)
*	*	*	*	*	*	*	*	1	#	#	0							4w3	(9.24)
*	*	*	*	*	*	*	*	0	0	*	*							5h(0) = 5h1	(9.25)
*	*	*	*	*	*	0	#	#	1	*	*							4o(−2)	(9.26)
*	*	*	*	*	*	#	0	#	1	*	*							4o(−1)	(9.27)
*	*	*	*	*	*	#	#	0	1	*	*							4o(0)	(9.28)
*	*	*	*	0	#	#	#	#	2	*	*							3o(−4)	(9.29)
*	*	*	*	#	0	#	#	#	2	*	*							3o(−3)	(9.30)
*	*	*	*	#	#	0	#	#	2	*	*							3o(−2)	(9.31)
*	*	*	*	#	#	#	0	#	2	*	*							3o(−1)	(9.32)
*	*	*	*	#	#	#	#	0	2	*	*							3o(0)	(9.33)
*	*	0	#	#	#	#	#	#	3	*	*							2o(−6)	(9.34)
*	*	#	0	#	#	#	#	#	3	*	*							2o(−5)	(9.35)
*	*	#	#	0	#	#	#	#	3	*	*							2o(−4)	(9.36)
*	*	#	#	#	0	#	#	#	3	*	*							2o(−3)	(9.37)
*	*	#	#	#	#	0	#	#	3	*	*							2o(−2)	(9.38)
*	*	#	#	#	#	#	0	#	3	*	*							2o(−1)	(9.39)
*	*	#	#	#	#	#	#	0	3	*	*							2o(0)	(9.40)
0	#	#	#	#	#	#	#	#	4	*	*							1o(−8)	(9.41)
#	0	#	#	#	#	#	#	#	4	*	*							1o(−7)	(9.42)
#	#	0	#	#	#	#	#	#	4	*	*							1o(−6)	(9.43)
#	#	#	0	#	#	#	#	#	4	*	*							1o(−5)	(9.44)
#	#	#	#	0	#	#	#	#	4	*	*							1o(−4)	(9.45)
#	#	#	#	#	0	#	#	#	4	*	*							1o(−3)	(9.46)
#	#	#	#	#	#	0	#	#	4	*	*							1o(−2)	(9.47)
#	#	#	#	#	#	#	0	#	4	*	*							1o(−1)	(9.48)
#	#	#	#	#	#	#	#	0	4	*	*							1o(0)	(9.49)

Comments

We discuss guesses about the w-, h-, and o-families

Discussion above assumes, for the o-family, IOM(−8;−8,3;3) correlates with nature.

People interpret Standard Model physics as suggesting the existence of leptoquarks. We are uncertain as to the extent leptoquark-mediated interactions could convert anti-quarks to quarks (or vice-versa). We are uncertain as to the extent leptoquark-mediated interactions could convert leptons to quarks (or vice-versa). We think Standard Model physics correlates with IOM(−2;−2,3;3).

The next items pertain.

Gss.9.1	O-family bosons for which (for the ground state) $n_{-2}=0$ or $n_{-1}=0$ transfer charge.	(9.50)
Gss.9.2	At least 1 $o(−2) particle has charge symbolized by Q'=−n/3, Here, 4≥$≥1. Here, n=1, 2, or 4. At least 1 $o(−1) particle has charge symbolized by Q'=+n/3. For each $, the charge of $o(−2) is the negative of the charge of $o(−1). Particles $o(0) have 0 charge.	(9.51)

For the remainder of this paper, we leave the matter of values for n unresolved.

The next items seem possible, but not certain.

Q' for \$o(−2) equals Q' for 4o(−2), for 3≥\$≥1 (9.52)
Q' for \$o(−1) equals Q' for 4o(−1), for 3≥\$≥1 (9.53)

The next items pertain.

Gss.9.3	O-family bosons for which (for the ground state) n_{-4}=0 or n_{-3}=0 transfer property-3.	(9.54)
Gss.9.4	O-family bosons for which (for the ground state) n_{-6}=0 or n_{-5}=0 transfer magnetic moment.	(9.55)
Gss.9.5	O-family bosons for which (for the ground state) n_{-8}=0 or n_{-7}=0 transfer property-1.	(9.56)
Gss.9.6	For the w-, h-, and o-families, closed oscillator pairs within the range IOM(−8;−8,3;3) correspond to channels. No other channels pertain.	(9.57)

We explore such matters further in Section 11 and Section 13.

We discuss aspects of the o-family

People say that quarks do not exist as free-ranging particles. We find Ω<0 for quarks. [Section 11] We reason by analogy. The next items apply.

Gss.9.7	Basic particles for which Ω<0 cannot range freely.	(9.58)
Gss.9.8	O-family basic particles are created in (at least) pairs or triplets.	(9.59)

We discuss binding energies for pairs or triplets of o-family particles

The mass of a pion exceeds the sum of the masses of its 2 constituent quarks. The mass of a nucleon, such as a proton, exceeds the sum of the masses of its 3 constituent quarks.

We do not know the extent to which property-3 for a pair or triplet of o-family particles varies from the sum of the property-3s of the constituent o-family members.

We discuss fields

Item (4.65) provides for twice as many solutions as we note above in this section. The next items provide an interpretation relevant to each pair s+1 and s−1 of solutions.

One linear combination of the 2 solutions correlates with a particle-creation operator (9.60)
The orthogonal combination of the 2 solutions correlates with a particle-destruction operator (9.61)

We suggest research

SOR.9.1	Determine the extent to which strengths of interactions mediated by w- , h-, and o-family basic bosons correlate with the concept of channels.
SOR.9.2	Verify or rule out (to some confidence level) existence of o-family bosons.
SOR.9.3	Determine properties (such as charge, mass, and magnetic moment) of o-family bosons.
SOR.9.4	Determine ranges for o-family forces.

SOR.9.5 Verify or rule out (to some confidence level) that o-family bosons cannot be created singly.
SOR.9.6 Determine or rule out (to some confidence level) non-zero binding energies for pairs and triplets of o-family bosons.
SOR.9.7 Determine the extent to which o-family bosons provide for the strong interaction's varying from R^0 spatial dependence.
SOR.9.8 Verify or rule out (to some confidence level) changes to nuclear theory people propose based on o-family physics.
STR.9.1 How best might people explore the existence and characteristics of $o% particles?
STR.9.2 What known or new phenomena people might explain based on the o-family?
STR.9.3 Estimate properties of o-family bosons.
STR.9.4 Estimate ranges for o-family forces.
STR.9.5 To what extent do o-family bosons correspond to aspects of the shell model for atomic nuclei? (Harmonic-oscillator math seems to pertain to each of the o-family and the shell model.)
STR.9.6 To what extent might people explain properties of atomic nuclei, based on o-family bosons (and gluons and other physics)?
STR.9.7 To what extent might people explain properties of neutron stars, based on o-family bosons (and other physics)?
STR.9.8 To what extent might people benefit by exploring the notion that o-family bosons erase and paint fermion properties? (Here, we have in mind possible parallels to people's considering that gluons erase and paint color charge.)
STR.9.9 Harmonize theory and observations or experiments regarding numbers of channels for interactions mediated by w-, h-, and o-family basic bosons.

Section 10 Gluons (the s-family)

Abs.10.1 S-family bosons provide for gluons for each of 2 sets of 3 color charges.

Context

We review matters related to the strong interaction

People state that the strong interaction binds quarks into mesons and into baryons. Examples of baryons include protons and neutrons. People state that gluons intermediate the strong interaction. People state that gluons have 0 mass. People state that the strong interaction asymptotically has R^0 spatial dependence. People state that color charge is a property associated with the strong interaction.

We correlate IOM with gluons

We anticipate describing an s-family and correlating the s-family with gluons.

Core

We list s-family basic bosons

The next items show s-family basic bosons. We use IOM(b_1;−2,1;b_4). The inclusion of the (−2)-and-(−1) oscillator pair correlates with S=1. Possibly, based on e-family results, b_1=−2 and b_4=9.

n_{-2}	n_{-1}	n_0	n_1	n_2	n_3	n_4	n_5	n_6	n_7	n_8	n_9	(10.1)
−2	−1	0	−2	*	*	*	*	*	*	*	*	(10.2)
0	−2	−1	−2	*	*	*	*	*	*	*	*	(10.3)
−1	0	−2	−2	*	*	*	*	*	*	*	*	(10.4)
−1	−2	0	−2	*	*	*	*	*	*	*	*	(10.5)
0	−1	−2	−2	*	*	*	*	*	*	*	*	(10.6)
−2	0	−1	−2	*	*	*	*	*	*	*	*	(10.7)

For each s-family basic boson, the next items pertain. Here we use numbers (not names of colors) for the 3 color charges. The numbers are −2, −1, 0 (as in n_{-2}, n_{-1}, and n_0, respectively).

For exactly 1 value of $\chi \leq 0$, $n_\chi = -2$ (10.8)
For exactly 1 value of $\chi' \leq 0$, $n_{\chi'} = 0$ (10.9)
Oscillator χ can be excited 1 time (10.10)

- Such an excitement includes the following
    - A quark loses color charge χ
    - n_χ becomes −1
    - $n_{\chi'}$ becomes −1

People might say that such an excitement erases color charge from a quark.

Similarly, a de-excitement from a state having $n_{-2} = n_{-1} = n_0 = -1$ paints the quark with a color charge.

In the next item, 1 trio consists of items (10.2), (10.3), and (10.4). Here, the cyclic order for quantum numbers is −2, −1, 0. Another trio consists of items (10.5), (10.6), and (10.7). Here, the cyclic order for quantum numbers is −2, 0, −1.

Gss.10.1 One trio of s-family bosons provides for gluons pertaining to quarks people consider to be matter. The other trio pertains to quarks people consider to be antimatter. (10.11)

We relate gluons and s-family basic bosons

For discussion, we assume that the first trio corresponds to quarks (and that the second trio corresponds to anti-quarks). (Possibly, the reversed pairing pertains.) The next item symbolizes a component for a gluon. The right element erases color charge −1. The left element paints color charge −2.

| item (10.2) > < item (10.3) | (10.12)

People sometimes denote the 3 color charges by r (for red), b (for blue), and g (for green). For such, we use ύ' to denote erasing color charge ύ. We use ó to denote painting color charge ó. For discussion, we assume color charge −1 (as in n_{-1}) corresponds to r and color charge −2 (as in n_{-2}) corresponds to b. The next item restates item (10.12).

br' (10.13)

The next items show 2 gluons of which item (10.13) comprises a component.

$$(rb' + br') / 2^{1/2} \quad (10.14)$$
$$-i(rb' - br') / 2^{1/2} \quad (10.15)$$

Item (10.16) provides a way people symbolize another 1 of the 8 gluons.

$$(rr' + bb' - 2gg') / 6^{1/2} \quad (10.16)$$

Perhaps, the standard (r, b, and g) representation aligns with our representation (−2 (as in n_{-2}), −1, and 0) via the following item.

g (in the r, b, g representation) correlates with 0 (in the −2, −1, and 0 representation) (10.17)

Comments

We note an interpretation

Presumably, people can consider that $\Omega=-2$ for s-family.

We note the matter of handedness/chirality

We think the totality of the next items correlates with handedness/chirality for quarks (and the opposite handedness for anti-quarks).

A correlating of W^- with a specific 1 of the 2-oscillator and the 3-oscillator (10.18)
A correlating of even-numbered oscillators with each other (and of odd numbered oscillators with each other) (10.19)
A correlating of a specific 1 of the 2 trios of s-family bosons with a specific 1 of quarks and anti-quarks (10.20)

We suggest research

SOR.10.1 Determine the number of channels that pertain for gluon-mediated (or s-family-meditated) interactions.

STR.10.1 Develop enough theory to enable experiments to determine the number of s-family channels.

STR.10.2 To what extent might people benefit by exploring the possibility that the spatial dependence of s-family forces is R^0 (asymptotic freedom) and that o-family (and possibly h-family) bosons provide for variation of the strong force from R^0 spatial dependence?

STR.10.3 To what extent might people find it appropriate to associate $(t')^{-2}$ behavior with 4s% interactions? (Here, t' denotes time.)

STR.10.4 To what extent might people benefit by considering the possibility that, if \$s(−4,−2) and \$s(−3,−1) coherences exist, the spatial character of the related force could be R^2? (Here, we have in mind a possible series - 1e2468& ↔ R^{-8}, ... , 4e2& ↔ R^{-2} , 4s% ↔ R^0, ...)

Section 11 Basic fermions (the q- and l-families)

Abs.11.1 IOM correlates with leptons, quarks, and related fields.
Abs.11.2 IOM correlates with possible basic fermions with S=3/2 and with S=7/2.
Abs.11.3 One IOM interpretation correlates with each n-type (or neutrino-like) basic fermion being its own antiparticle. One IOM interpretation correlates with each basic fermion being distinct from its antiparticle.
Abs.11.4 Each q- or l-family particle is a member of a 3-generation trio.

Context

We discuss people's lack of correlating harmonic oscillators with fermions

As far as we know, people tend to correlate harmonic oscillators with bosons and not with fermions.
As far as we know, people tend to focus on harmonic-oscillator solutions for which 2ν is an even integer. As far as we know, people tend to underutilize solutions for which 2ν is an odd integer.

We discuss a possible basis for correlating harmonic oscillators with fermions

IOM solutions for which 2ν is an odd integer exist. We anticipate correlating such odd-integer solutions with aspects of fermions. We focus on interactions fermions have with bosons.

Core

We define n-type

The next item provides a term for some q- and l-family basic particles. Among S=1/2 basic fermions, n-type correlates with neutrinos.

N-type denotes any q- or l-type basic fermion that (11.1)
- Has zero charge
- Can absorb charge via interactions with a 4w2 (W^- boson), 4w3 (W^+ boson), \$o(−2) boson, or \$o(−1) boson

We set a scope for the l- and q-families

We use QM-type-CS. We correlate edge solutions with particles. We correlate inside solutions with fields. Based on items (4.35) and (5.17), the next items pertain for basic fermions.

D for fields = (5 − 4Ω) / 2 (11.2)
D for particles = (21 − 4Ω) / 6 (11.3)

Based on known numbers of particles in the q- and l-families, 1 of the next 2 items pertains. The 2(2S+1) choice correlates with each particle having a distinct antiparticle. We call this choice n-type-S. The D choice correlates with n-type particles being their own antiparticles. We call this choice n-type-D.

Number of particles per generation = 2(2S+1) for particles (11.4)
Number of particles per generation = D for particles (11.5)

Based on item (4.68), the next item pertains.

Number of particles = 3 × (number of particles per generation) (11.6)

The next items show results. Rows for which both the traditional particles column and the possible particles column show blanks fall outside assumptions we make. Each # per gen column denotes number of particles per generation. The left one ("S") of those 2 columns correlates with item (11.4) and with the term n-type-S. The right one (D) of those 2 columns correlates with item (11.5) and with the term n-type-D.

S	Ω	D for fields	D for particles	Traditional particles	Possible particles	# per gen "S"	# per gen D	(11.7) (11.8)
3/2	15/4	−5	1					(11.9)
1/2	3/4	1	3	Leptons		4	3	(11.10)
1/2	−3/4	4	4	Quarks		4	4	(11.11)
3/2	−15/4	10	6		q(3/2)	8	6	(11.12)
5/2	−35/4	20	28/3					(11.13)
7/2	−63/4	34	14		q(7/2)	16	14	(11.14)
...	−S(S+1)							(11.15)

We discuss IOM for l-family particles and S=1/2 q-family particles

The next items apply for lepton particles and for quark particles. Here, L denotes a principal quantum number for spin-like systems that correlate with $-r_e \leq r \leq 0$. Here, M denotes a secondary quantum number. We use results from items that start with item (4.64).

For each 1 of 3 generations, the following number of relevant solutions correlates with each of S=1/2 with Ω=+3/4 and S=1/2 with Ω=−3/4 (11.16)

- 2(2S+1) = 4

Gss.11.1 For the l-family, combinations of the 4 solutions correspond to 2 of the 3 possible members of an L=1 set (the M=0 member does not apply) and to the 1 member (M=0) of an L=0 set. (11.17)

Gss.11.2 For the q-family, for S=1/2, combinations of the 4 solutions correspond to 4 of the 5 members of an L=2 set (the M=0 member does not apply). (11.18)

We interpret items following item (11.19) as correlating with and supporting items (11.17) and (11.18). In the next items, M" and M' are integer indices. The 0≤M"≤3 rows show an orderly array of particle masses for charged leptons and quarks. Here, M' correlates with (but does not necessarily equal) M. For each particle, an item shows $\log_{10}(\text{mass}/m_e)$, charge in units of Q', and particle name.

M" \ M'	−3	−2	−1	0	1	...	(11.19) (11.20)
0	0.00 (−1) electron	0.61 (+2/3) up	0.97 (−1/3) down		0.97 (+1/3) anti-down		(11.21)

	M'	−3	−2	−1	0	1	…	(11.19)
M"								(11.20)
1			2.26 (−1/3) strange	3.40 (+2/3) charm		3.40 (−2/3) anti-charm		(11.22)
2		2.32 (−1) muon	3.93 (−1/3) bottom	5.51 (+2/3) top		5.51 (−2/3) anti-top		(11.23)
3		3.54 (−1) tauon						(11.24)

The next items correlate item (11.17), item (11.18), and items following item (11.19). Here, we consider all 3 generations.

For each of the rows for which M"=0, 2, or 3, the following apply (11.25)

- The row includes an instance characterized by L=1
  - M' = −3, 0, and +3 characterize this instance
  - A lepton and anti-lepton pair correlate with M' = −3 and M' = +3
  - No particle correlates with M'=0 for that row
- A corresponding L = 0, M=0 row has M" ≤ −3

For each of the rows for which M"=0, 1, or 2, the following apply (11.26)

- The row includes an instance characterized by L=2
  - M' = −2, −1, 0, +1, and +2 characterize this instance
  - A quark and anti-quark pair correlate with M' = −2 and M' = +2
  - A quark and anti-quark pair correlate with M' = −1 and M' = +1
  - No particle correlates with M'=0

For each of the rows for which M"=0 or 2, the following apply (11.27)

- The row includes an instance characterized by L=3
  - M' = −3, −2, −1, 0, +1, +2, and +3 characterize this instance
  - A lepton and anti-lepton pair correlate with M' = −3 and M' = +3
  - A quark and anti-quark pair correlate with M' = −2 and M' = +2
  - A quark and anti-quark pair correlate with M' = −1 and M' = +1
  - No particle correlates with M'=0

The next item applies.

Gss.11.3 For χ an even positive integer, each of the | n_χ , $n_{\chi+1}$ > states denoted by | −1 , −2 > or by | −2 , −1 > corresponds to spin/ħ = 1/2. (11.28)

The next items show a technique for representing charged leptons (l-family) and quarks (the S=1/2 part of the q-family). We use IOM(−8;−2,3;9). We describe each particle's possibility for absorbing a property from an interaction with a w-, h-, or o-family boson. For each of these items, Œ=0. Here, we discuss generation-1 particles. For each oscillator χ for which we show n_χ=−2, absorption of one unit of property can occur. If such χ is −2 or −1, a change in the fermion's charge occurs. If such χ is 0 or 1, a change of generation occurs. If such χ is 2 or 3, |Q'|=1 provides the magnitude of the change in charge.

n_{-8}	n_{-7}	n_{-6}	n_{-5}	n_{-4}	n_{-3}	n_{-2}	n_{-1}	n_0	n_1	n_2	n_3	n_4	n_5	n_6	n_7	n_8	n_9	Particles	(11.29)
*	*	*	*	*	*	-1	-1	-2	-1	-2	-1	*	*	*	*	*	*	positron	(11.30)
*	*	*	*	*	*	-1	-1	-2	-1	-1	-2	*	*	*	*	*	*	electron	(11.31)
*	*	*	*	*	*	-2	-1	-2	-2	-2	-1	*	*	*	*	*	*	up	(11.32)
*	*	*	*	*	*	-1	-2	-2	-2	-2	-1	*	*	*	*	*	*	anti-down	(11.33)
*	*	*	*	*	*	-2	-1	-2	-2	-1	-2	*	*	*	*	*	*	down	(11.34)
*	*	*	*	*	*	-1	-2	-2	-2	-1	-2	*	*	*	*	*	*	anti-up	(11.35)

Items (11.30) and (11.31) correlate with 2 of the 4 l-family solutions. Other items following item (11.29) correlate with 4 of the 4 q-family S=1/2 solutions.

The next items correlate with the other 2 l-family solutions. For each of these items, Œ=0. For n-type-S, each of neutrino-a and neutrino-b correlates with a neutrino. For n-type-D, a linear combination of the items labeled neutrino-a and neutrino-b correlates with a neutrino. Presumably, for n-type-D, the orthogonal linear combination does not represent a particle.

n_{-8}	n_{-7}	n_{-6}	n_{-5}	n_{-4}	n_{-3}	n_{-2}	n_{-1}	n_0	n_1	n_2	n_3	n_4	n_5	n_6	n_7	n_8	n_9	Concept	(11.36)
*	*	*	*	*	*	-1	-1	-2	-1	-2	-1	*	*	*	*	*	*	(neutrino-a)	(11.37)
*	*	*	*	*	*	-1	-1	-2	-1	-1	-2	*	*	*	*	*	*	(neutrino-b)	(11.38)

These representations correlate with quarks being able to change generations via Z bosons. These representations correlate with leptons not changing generation via Z bosons.

We explore IOM for q-family particles with S=3/2

We interpret items following item (11.7) as correlating with possibilities for q-family basic fermions with S=3/2 and S=7/2.

The next item correlates with S=3/2.

Gss.11.4 Q-family basic fermions for which S=3/2 have either n_4=−1 and n_5=−2 or n_4=−2 and n_5=−1. (11.39)

Each particle listed below would have 3 generations.

The next items correlate with the first generations of each of 4 basic particles we associate with an L=2 set. In the notation q(3/2;ї,ύ), ї correlates with $n_{ї}$=−2 and ύ correlates with $n_{ύ}$=−2. Here, ї≤−1 and ύ≥4.

n_{-8}	n_{-7}	n_{-6}	n_{-5}	n_{-4}	n_{-3}	n_{-2}	n_{-1}	n_0	n_1	n_2	n_3	n_4	n_5	n_6	n_7	n_8	n_9	Particles	(11.40)
*	*	*	*	-2	-1	-1	-1	-2	-2	-1	-1	-2	-1	*	*	*	*	q(3/2;−4,4)	(11.41)
*	*	*	*	-1	-2	-1	-1	-2	-2	-1	-1	-2	-1	*	*	*	*	q(3/2;−3,4)	(11.42)
*	*	*	*	-2	-1	-1	-1	-2	-2	-1	-1	-1	-2	*	*	*	*	q(3/2;−4,5)	(11.43)
*	*	*	*	-1	-2	-1	-1	-2	-2	-1	-1	-1	-2	*	*	*	*	q(3/2;−3,5)	(11.44)

Based on $n_{ύ}$=−1 for ύ = −2, −1, 2, and 3, these S=3/2 particles cannot interact with charged w-family bosons or with charged o-family bosons.

The next items correlate with S=3/2 n-type particles. For n-type-S, 4 particles exist. For n-type-D, a linear combination of item (11.46) and item (11.47) correlates with 1 particle and a linear combination of item (11.48) and item (11.49) correlates with another particle.

n_{-8}	n_{-7}	n_{-6}	n_{-5}	n_{-4}	n_{-3}	n_{-2}	n_{-1}	n_0	n_1	n_2	n_3	n_4	n_5	n_6	n_7	n_8	n_9	Concept	(11.45)
*	*	*	*	−1	−1	−2	−1	−2	−2	−1	−1	−2	−1	*	*	*	*	qa(3/2;−2,4)	(11.46)
*	*	*	*	−1	−1	−1	−2	−2	−2	−1	−1	−2	−1	*	*	*	*	qb(3/2;−1,4)	(11.47)
*	*	*	*	−1	−1	−2	−1	−2	−2	−1	−1	−1	−2	*	*	*	*	qa(3/2;−2,5)	(11.48)
*	*	*	*	−1	−1	−1	−2	−2	−2	−1	−1	−1	−2	*	*	*	*	qb(3/2;−1,5)	(11.49)

For these n-type particles, interactions with o-family charged particles can occur. For these n-type particles, no interactions with W bosons occur. Also, each of the n-type particles has no net charge.

The next items pertain.

Each S=3/2 basic fermion has no charge (that is, Q'=0) (11.50)
Each S=3/2 n-type basic fermion can absorb charge (11.51)

The next item pertains.

Gss.11.5 If an S=3/2 basic fermion absorbs charge, the fermion becomes a quark or a charged lepton. (11.52)

The next item reviews concepts correlating with S=3/2 fermions.

For S=3/2, for each of 3 generations, the 8=2(2S+1) solutions correlate with an L=2 set and 2 L=0 sets (11.53)

- The L=2 set correlates with 4 particles
    - This set lacks its M=0 member
- Each L=0 set correlates with 2 particles (if n-type-S applies) or correlates with 1 particle (if n-type-D applies)
    - Here, M=0

We explore IOM for q-family particles with S=7/2

The next items correlate with the first generations of 12 basic particles we associate with S=7/2.

n_{-8}	n_{-7}	n_{-6}	n_{-5}	n_{-4}	n_{-3}	n_{-2}	n_{-1}	n_0	n_1	n_2	n_3	n_4	n_5	n_6	n_7	n_8	n_9	Particles	(11.54)
−2	−1	−1	−1	−1	−1	−1	−1	−2	−2	−1	−1	−1	−1	−1	−1	−2	−1	q(7/2;−8,8)	(11.55)
−1	−2	−1	−1	−1	−1	−1	−1	−2	−2	−1	−1	−1	−1	−1	−1	−2	−1	q(7/2;−7,8)	(11.56)
−2	−1	−1	−1	−1	−1	−1	−1	−2	−2	−1	−1	−1	−1	−1	−1	−1	−2	q(7/2;−8,9)	(11.57)
−1	−2	−1	−1	−1	−1	−1	−1	−2	−2	−1	−1	−1	−1	−1	−1	−1	−2	q(7/2;−7,9)	(11.58)
−1	−1	−2	−1	−1	−1	−1	−1	−2	−2	−1	−1	−1	−1	−1	−1	−2	−1	q(7/2;−6,8)	(11.59)
−1	−1	−1	−2	−1	−1	−1	−1	−2	−2	−1	−1	−1	−1	−1	−1	−2	−1	q(7/2;−5,8)	(11.60)
−1	−1	−2	−1	−1	−1	−1	−1	−2	−2	−1	−1	−1	−1	−1	−1	−1	−2	q(7/2;−6,9)	(11.61)
−1	−1	−1	−2	−1	−1	−1	−1	−2	−2	−1	−1	−1	−1	−1	−1	−1	−2	q(7/2;−5,9)	(11.62)
−1	−1	−1	−1	−2	−1	−1	−1	−2	−2	−1	−1	−1	−1	−1	−1	−2	−1	q(7/2;−4,8)	(11.63)
−1	−1	−1	−1	−1	−2	−1	−1	−2	−2	−1	−1	−1	−1	−1	−1	−2	−1	q(7/2;−3,8)	(11.64)
−1	−1	−1	−1	−2	−1	−1	−1	−2	−2	−1	−1	−1	−1	−1	−1	−1	−2	q(7/2;−4.9)	(11.65)
−1	−1	−1	−1	−1	−2	−1	−1	−2	−2	−1	−1	−1	−1	−1	−1	−1	−2	q(7/2;−3,9)	(11.66)

The next items pertain to S=7/2 n-type particles. For n-type-S, 4 particles exist. For n-type-D, a linear combination of item (11.68) and item (11.69) correlates with 1 particle and a linear combination of item (11.70) and item (11.71) correlates with another particle.

n_{-8}	n_{-7}	n_{-6}	n_{-5}	n_{-4}	n_{-3}	n_{-2}	n_{-1}	n_0	n_1	n_2	n_3	n_4	n_5	n_6	n_7	n_8	n_9	Concept	(11.67)
−1	−1	−1	−1	−1	−1	−2	−1	−2	−2	−1	−1	−1	−1	−1	−1	−2	−1	qa(3/2;−2,8)	(11.68)
−1	−1	−1	−1	−1	−1	−1	−2	−2	−2	−1	−1	−1	−1	−1	−1	−2	−1	qb(3/2;−1,8)	(11.69)
−1	−1	−1	−1	−1	−1	−2	−1	−2	−2	−1	−1	−1	−1	−1	−1	−1	−2	qa(3/2;−2,9)	(11.70)
−1	−1	−1	−1	−1	−1	−1	−2	−2	−2	−1	−1	−1	−1	−1	−1	−1	−2	qb(3/2;−1,9)	(11.71)

The next item reviews concepts correlating with S=7/2 fermions.

For S=7/2, for each of 3 generations, the 16=2(2S+1) solutions correlate with 3 L=2 sets and 2 L=0 sets (11.72)

- Each L=2 set correlates with 4 particles
  - Each such set lacks its M=0 member
- Each L=0 set correlates with 2 particles (if n-type-S applies) or correlates with 1 particle (if n-type-D applies)
  - Here, M=0

The next items pertain.

Each S=7/2 basic fermion has no charge (that is, Q'=0) (11.73)
Each S=7/2 n-type basic fermion can absorb charge (11.74)

The next item pertains.

Gss.11.6 If an S=7/2 basic fermion absorbs charge, the fermion becomes a quark or a charged lepton. (11.75)

Comments

We note that IOM may correlate with fermion magnetic monopoles not existing

The next item may pertain, based on items including and following item (6.25) and on item (11.13).

Fermion magnetic monopoles would correlate with S=5/2 and do not exist (11.76)

We discuss fermion fields

We note items (4.66) and (4.68). For a given combination of S and Ω, as many field-oriented solution sets exist as do particle-oriented solution sets. The number of field-oriented solutions is 2(2S+1). For particles that correlate with L≠0 sets, pairing a particle and an antiparticle pairs 2 field solutions. We assume 1 linear combination of the 2 field solutions correlates with creating a particle-antiparticle pair. The orthogonal linear combination of the 2 field solutions correlates with destroying a particle-antiparticle pair. For n-type-S, similar considerations apply. For n-type-D, similar considerations possibly apply to the components for particles that correlate with L=0 sets.

The next items pertain, based on comparing the number of field-correlated solutions and the number of particle-correlated solutions.

Fermion fields do not correlate with generation	(11.77)
Pair creation (or pair annihilation) that correlates with fields does not correlate with generation	(11.78)

We suggest research

SOR.11.1 Detect or rule out (to some confidence level) the existence of basic fermions for which S=3/2 or S=7/2.

SOR.11.2 Rule out (to some confidence level) or detect the existence of basic fermions for which S=5/2.

SOR.11.3 Measure or infer properties of S≥3/2 basic fermions.

SOR.11.4 Detect or rule out (to some confidence level) the existence of interactions that convert fermions between S≥3/2 n-type and S=1/2.

SOR.11.5 Verify or rule out (to some confidence level) q- and l-family interaction rules we show regarding the w-, h-, and o-families. Determine strengths for interactions for which strengths are yet to be determined.

SOR.11.6 To what extent does either n-type model (n-type-S or n-type-D) better correlate with observations than does the other n-type model?

STR.11.1 Estimate masses and magnetic moments for basic fermions for which S≥3/2.

STR.11.2 Describe possible composite objects (such as nuclei or atoms) for which S≥3/2 basic fermions would be components.

Part 4 Particles and phenomena

Context

We discuss some unresolved physics

People say that people do not adequately understand mechanisms governing changes in the observed rate of expansion of the universe. People say that models do not adequately correlate with the observed universe's having much more matter than antimatter. People say that the Standard Model does not adequately correlate with observed sizes of violations of symmetries related to CPT (charge, parity, and time) symmetry. People do not know masses for neutrinos. People say that models do not adequately correlate with neutrino oscillations. Perhaps, people would say that the Standard Model does not adequately correlate with masses of elementary particles.

We anticipate addressing such unresolved matters

We anticipate suggesting aspects of IOM that correlate with resolving such matters.

Core

We preview sections in this part

One section correlates changes in the rate of expansion of the universe with e-family coherences.

One section addresses some topics people say the Standard Model does not adequately address. We provide a mechanism leading to matter/antimatter imbalance. We discuss CPT-related symmetry violations and provide a basis for people's not being able to explain such violations fully via the Standard Model.

One section derives $E^2-c^2P^2=0$ for the e-family.

One section provides an approximate formula linking masses of non-zero-mass basic bosons.

One section provides an approximate formula for masses of non-zero-charge basic fermions.

One section estimates strengths for various e-family-mediated interactions.

One section discusses mechanisms for various phenomena, including neutrino oscillations.

Section 12 The rate of expansion of the universe

Abs.12.1 E-family coherences provide for changes in the rate of expansion of the universe.

Context

We review astrophysical observations about the rate of expansion of baryonic matter

Traditional physics includes observations indicating that directly observed astrophysical objects move away from each other. People discuss a concept of a rate of expansion of the universe. People say that the rate varies during the history of the universe.

People deduce rates of expansion from observations of photons. People express approximate time after the big bang in terms of the redshift, z, people attribute to photons people and equipment observe.

Redshift z=0 pertains to photons emitted recently. Redshift numbers are positive. Any value of z denotes a time before the time people associate with a smaller value of z.

These items pertain. 3 eras exist. [Ref.12.1 and Ref.12.2]

For $z >$ some number greater than 2.3, the observed rate of expansion increases • We call this era Z1	(12.1)
During $\sim 2.3 > z > \sim 0.7$, the observed rate of expansion decreases • We call this era Z2	(12.2)
During $0.46 \pm 0.13 > z > 0$, the observed rate of expansion increases • We call this era Z3	(12.3)

Traditional physics does not seem to include interactions that could cause such changes in the rate of expansion. Sometimes, people seem to correlate increasing rate of expansion with dark energy.

We anticipate correlating 3 periods in the expansion with 3 e-family coherences

Above, we suggest that the 1e2468& coherence provides an R^{-8} force. The 2e246& coherence provides the likely most important R^{-6} force. The 3e24& coherence provides the likely most important R^{-4} force. We anticipate correlating these forces with eras in the evolution of the baryonic matter.

Core

We posit mechanisms affecting the rate of expansion

We think about 2 clumps of stuff. These clumps have similar size. The clumps neighbor each other.

The next items discuss forces that dominate within each clump and between the 2 clumps. These items reflect the observation that the universe expands. During such expansion, effects of an $R^{-2ú}$ force (within or between the 2 clumps) evolve from being more than effects of $R^{-2(ú-1)}$ forces to being less than effects of $R^{-2(ú-1)}$ forces.

During era E1, the force associated with 1e2468& dominates	(12.4)
During era E2, the force associated with 2e246& dominates	(12.5)
During era E3, the force associated with 3e24& dominates	(12.6)
During era E4, the forces associated with 3e4& and 4e2& dominate	(12.7)

Today, observed atoms, planets, stars, galaxies, and galactic clusters exhibit era E4 behavior. At a size larger than galactic superclusters, people observe E3 behavior.

The next items pertain.

Gss.12.1	For observed astrophysical objects of above some size, era Eó correlates with era Zú, for $1 \le ó = ú \le 3$.	(12.8)
Gss.12.2	1e2468& repels astrophysical objects from each other. 2e246& attracts astrophysical objects to each other. 3e24& repels astrophysical objects from each other.	(12.9)

Comments

We note possibilities that E2 behavior still dominates for some objects

Should the universe be adequately vast, large objects could still experience E2 behavior.

We suggest research

SOR.12.1 Estimate strengths and directions (attraction or repulsion) for e-family forces other than 4e2& and 3e4&.
SOR.12.2 Estimate charges of objects for which 3e24& currently dominates.
STR.12.1 Predict strengths and directions (attraction or repulsion) for e-family forces other than 4e2& and 3e4&.
STR.12.2 To what extent do neutrinos interact with $e%&-for-which-2ϵ% bosons based on, for example, neutrinos being transformed into virtual pairs, each consisting of a charged lepton and a 4w2 or 4w3?

We list references

Ref.12.1 N. G. Busca, et. al., Baryon Oscillations in the Lyα forest of BOSS quasars, arXiv:1211.2616 [astro-ph.CO].
Ref.12.2 A. Riess, et. al., Type Ia Supernova Discoveries at z > 1 from the *Hubble Space Telescope*: Evidence for Past Deceleration and Constraints on Dark Energy Evolution, *The Astrophysical Journal*, 607, 665 (2004). (doi:10.1086/383612) (http://iopscience.iop.org/0004-637X/607/2/665)

Section 13 Matter/antimatter imbalance and CPT-related symmetries

Abs.13.1 Lasing of o-family particles provided key effects leading to matter/antimatter imbalance.
Abs.13.2 O-family particles provide for CPT-related symmetry violations.
Abs.13.3 O-family particles with S≥2 close gaps between magnitudes of violations people estimate via the Standard Model and magnitudes people measure.
Abs.13.4 E-family coherences provide for phenomena people attribute to axions.

Context

We discuss various asymmetries

People speculate as to why baryonic matter has much more matter (such as electrons and protons) than antimatter (such as positrons and antiprotons). People speculate as to sources of observed symmetry violations, for example with respect P (parity) or CP (charge and parity). People say that the Standard Model can account for some P violation and CP violation. People say that the Standard Model does not account for observed amounts of P violation and CP violation.

We anticipate providing explanations regarding asymmetries

Here, we provide possible explanations. We discuss o- and e-family phenomena with which the Standard Model does not correlate.

Core

We discuss matter/anti-matter imbalance

The next items show possible examples of interaction vertices. Such interactions would convert quarks into anti-quarks or anti-quarks into quarks.

Interaction vertices		|Q'|, for $o%	(13.1)
up + $o(−2) → anti-down	for some $ with 4≥$≥1	1/3	(13.2)
anti-up + $o(−1) → down	for some $ with 4≥$≥1	1/3	(13.3)
anti-down + $o(−2) → down	for some $ with 4≥$≥1	2/3	(13.4)
anti-up + $o(−1) → up	for some $ with 4≥$≥1	4/3	(13.5)

The next item shows a reaction that would convert a positron and the basic q-family components for an anti-proton into an electron and the basic q-family components for a proton.

1 Positron + 2 Anti-up + 1 Anti-down → 1 Electron + 2 Up + 1 Down (13.6)

The next items show reactions that would net to item (13.6). Here, each entry shows a value of Q'. Here, a fermion name with its first letter capitalized represents a fermion named in item (13.6). A fermion name with its first letter in lower case represents a fermion that the reactions create and then destroy. A +1 boson corresponds to a W⁺ or possibly a $o(−1) having Q'=+1. A −1 boson corresponds to a W⁻ or possibly a $o(−2) having Q'=−1. A +1/3 boson corresponds to a $o(−1) having Q'=+1/3.

This fermion	emits or absorbs this boson	and becomes this fermion.	This fermion	absorbs or emits this boson	and becomes this fermion.	(13.7)
−2/3 Anti-up	absorbs +1	+1/3 anti-down	+1 Positron	emits +1	0 neutrino	(13.8)
−2/3 Anti-up	emits −1	+1/3 anti-down	0 neutrino	absorbs −1	−1 Electron	(13.9)
+1/3 Anti-down	absorbs +1/3	+2/3 up	+2/3 up	emits +1	−1/3 Down	(13.10)
+1/3 anti-down	absorbs +1/3	+2/3 Up				(13.11)
+1/3 anti-down	absorbs +1/3	+2/3 Up				(13.12)

These reactions seem to imply that reactions such as at least 1 of the next items shows can occur. Here, we consider a possible $o(−1) having Q'=+1/3. Here, we consider a possible $"o(−1) having Q'=+1.

? + W⁺ = ? + 4w3 → 3 $o(−1) + ?', for some ?, some $ with 4≥$≥1, and some ?' (13.13)

? + $"o(−1) → 3 $o(−1) + ?',	for some ?, some $" with 4≥$"≥1, some $ with 4≥$≥1, and some ?'	(13.14)

The next items show reactions that would net to item (13.6). A +2/3 boson corresponds to a $o(−1) having Q'=+2/3.

This fermion	emits or absorbs this boson	and becomes this fermion.	This fermion	absorbs or emits this boson	and becomes this fermion.	(13.15)
−2/3 Anti-up	absorbs +1	+1/3 anti-down	+1 Positron	emits +1	0 neutrino	(13.16)
−2/3 Anti-up	emits −1	+1/3 anti-down	0 neutrino	absorbs −1	−1 Electron	(13.17)
+1/3 Anti-down	emits +2/3	−1/3 Down				(13.18)
+1/3 anti-down	emits +2/3	−1/3 down	−2/3 down	absorbs +1	+2/3 Up	(13.19)
+1/3 anti-down	emits +2/3	−1/3 down	−2/3 down	absorbs +1	+2/3 Up	(13.20)

These reactions seem to imply that reactions such as at least 1 of the next items shows can occur. Here, we consider a possible $o(−1) having Q'=+2/3. Here, we consider a possible $"o(−1) having Q'=+1.

? + 3 $o(−1) → 2 4w3 + ? '= 2 W^+ + ?',	for some ?, some $ with 4≥$≥1, and some ?'	(13.21)
? + 3 $o(−1) → 2 $"o(−1) + ?',	for some ?, some $ with 4≥$≥1, some $" with 4≥$"≥1, and some ?'	(13.22)

We could include a discussion (similar to 2 discussion just above) based on converting anti-ups to ups. For such, a +4/3 boson corresponds to a $o(−1) having Q'=+4/3.

Presumably, during part of the big bang, baryonic matter was sufficiently dense that many basic q-family particles interacted via $o%. A non-uniformity (in, for example, spatial distributions of quarks) or an instability could have led to lasing by $o% bosons. Presumably, the ? and ?' in item (13.13) or (13.14) or in item (13.21) or (13.22) could include e-family particles, which also could have lased.

The next item correlates with such reactions leading to matter/antimatter imbalances. The next item correlates with item (9.51).

Gss.13.1 For some $ (with 4≥$≥1) and some ύ (with ύ=1, 2, or 4), Q'=−ύ/3 for $o(−2) and Q'=+ύ/3 for $o(−1). (13.23)

The next items list imbalances such reactions would create.

Number of Q'=−1 leptons ≫ number of Q'=+1 (anti-)leptons (13.24)

Number of Q'=+2/3 quarks ≫ number of Q'=−2/3 (anti-)quarks (13.25)
Number of Q'=−1/3 quarks ≫ number of Q'=+1/3 (anti-)quarks (13.26)

We discuss an example of P-symmetry violation

The next item pertains.

Gss.13.2 People consider interactions that convert anti-quarks into quarks (or vice-versa) to violate P symmetry. (13.27)

The next item pertains. We base such on reactions such as items following item (13.1) indicate.

Gss.13.3 O-family bosons ($o(−2) and $o(−1), for some 4≥$≥1) mediate interactions that exhibit P violation. (13.28)

We discuss CP violation and the possibility for axions

People say that CP violations correlating with the Standard Model may be too small to correlate with observed magnitudes of CP violations. People note a concept of axions. People suggest axions might provide for some CP violation. People suggest that axions have some (small) mass and may decay into photons. We suggest the following.

Gss.13.4 Some o-family members $o% for which 4≥$≥1 and %=−2 or −1 contribute to CP violation approximately the amounts for which the Standard Model can account. (13.29)
Gss.13.5 Other o-family members contribute to CP violation, beyond that for which the Standard Model can account. (13.30)
Gss.13.6 Phenomena people associate with axions exist. People can associate such phenomena with e-family coherences. (13.31)

Items (13.29) and (13.30) correlate with the concept that the Standard Model correlates with IOM(−2;−2,3;3). [item (23.6)]

Item (13.31) correlates with the concept that the Standard Model correlates with IOM(−2;−2,3;3). [item (23.6)] E-family member 3e24& becomes a candidate for participating in such phenomena. Here, the relevant interaction would involve the graviton component of a 3e24% and would leave the photon component as a product. Here, the using of the graviton component could possibly correspond to traditional thoughts that people express as non-zero mass for axions. Possibly, any $e%& for which 2∊% and for which % has more than 2 elements contributes to phenomena people associate with axions.

We reinterpret traditional P-, C-, and T-symmetries and related violations

The next items describe some types of representations within IOM(−8;−8,9;9).

Families	b_2 for IOM(−8;−8,9;9)	b_3 for IOM(−8;−8,9;9)	(13.32)
e- and s-	−2	9	(13.33)
w-, h-, and o-	−8	3	(13.34)
l- and q-	−8	9	(13.35)

The next items show similar representations for IOM(−2;−2,3;3), which we think correlates with the Standard Model.

Families	b_2 for IOM(−2;−2,3;3)	b_3 for IOM(−2;−2,3;3)	
			(13.36)
e- and s-	−2	3	(13.37)
w-, h-, and o-	−2	3	(13.38)
l- and q-	−2	3	(13.39)

People traditionally discuss P-, C-, and T-symmetries based on physics that items following item (13.36) model.

The next items describe aspects of those symmetries. Here, the $p_{\acute{o}}$ refer to components of a 4-momentum 4 vector, with p_0 corresponding to the energy-related component. Otherwise, here, for ï↔ó, ï and ó refer to oscillator numbers.

Families	C-symmetry exchanges	P-symmetry exchanges	T-symmetry exchanges	(13.40)
e- and s-		$p_{\acute{\upsilon}} \leftrightarrow -p_{\acute{\upsilon}}$ $1 \leq \acute{\upsilon} \leq 3$	$p_0 \leftrightarrow -p_0$	(13.41)
w-, h-, and o-	$-2 \leftrightarrow -1$ $2 \leftrightarrow 3$			(13.42)
l- and q-	$-2 \leftrightarrow -1$ $2 \leftrightarrow 3$			(13.43)

IOM provides that n_1 aligns (for the e-family) with motion. The next items pertain for IOM(−2;−2,3;3).

Families	C-symmetry exchanges	P-symmetry exchanges	T-symmetry exchanges	(13.44)
e- and s-	$-2 \leftrightarrow -1$ $2 \leftrightarrow 3$	$2 \leftrightarrow 3$	$-2 \leftrightarrow -1$	(13.45)
w-, h-, and o-	$-2 \leftrightarrow -1$ $2 \leftrightarrow 3$	$2 \leftrightarrow 3$	$-2 \leftrightarrow -1$	(13.46)
l- and q-	$-2 \leftrightarrow -1$ $2 \leftrightarrow 3$	$2 \leftrightarrow 3$	$-2 \leftrightarrow -1$	(13.47)

Items following item (13.44) exhibit CPT symmetry. People state that CPT symmetry applies to Standard Model physics.

The next items pertain for IOM(−8;−8,9;9). Oscillators numbered −8 through −3 do not pertain to the e- and s-families. Oscillators numbered 4 through 9 do not pertain to the w-, h-, and o-families.

Families	C-symmetry exchanges	P-symmetry exchanges	T-symmetry exchanges	(13.48)
e- and s-	$-2 \leftrightarrow -1$ $2 \leftrightarrow 3$ $4 \leftrightarrow 5$ $6 \leftrightarrow 7$ $8 \leftrightarrow 9$	$2 \leftrightarrow 3$ $4 \leftrightarrow 5$ $6 \leftrightarrow 7$ $8 \leftrightarrow 9$	$-2 \leftrightarrow -1$	(13.49)

Families	C-symmetry exchanges	P-symmetry exchanges	T-symmetry exchanges	(13.48)
w-, h-, and o-	$-8 \leftrightarrow -7$	$2 \leftrightarrow 3$	$-8 \leftrightarrow -7$	(13.50)
	$-6 \leftrightarrow -5$		$-6 \leftrightarrow -5$	
	$-4 \leftrightarrow -3$		$-4 \leftrightarrow -3$	
	$-2 \leftrightarrow -1$		$-2 \leftrightarrow -1$	
	$2 \leftrightarrow 3$			
l- and q-	$-8 \leftrightarrow -7$	$2 \leftrightarrow 3$	$-8 \leftrightarrow -7$	(13.51)
	$-6 \leftrightarrow -5$	$4 \leftrightarrow 5$	$-6 \leftrightarrow -5$	
	$-4 \leftrightarrow -3$	$6 \leftrightarrow 7$	$-4 \leftrightarrow -3$	
	$-2 \leftrightarrow -1$	$8 \leftrightarrow 9$	$-2 \leftrightarrow -1$	
	$2 \leftrightarrow 3$			
	$4 \leftrightarrow 5$			
	$6 \leftrightarrow 7$			
	$8 \leftrightarrow 9$			

The items following item (13.48) exhibit CPT symmetry. The next item pertains.

Gss.13.7 Differences between ύ-symmetry for IOM(−8;−8,9;9) and ύ-symmetry for IOM(−2;−2,3;3) correlate with sizes of ύ-violations people do not associate with Standard Model physics. Here, ύ can be (at least) C, P, CP, or T. (13.52)

We discuss o-family charges

Item (9.51) suffices for work in this section regarding o-family charges.

Comments

We bring together some concepts regarding rotational order regarding n_{-2}, n_{-1}, and n_0

Section 10 notes a pair of distinct rotational orders for n_{-2}-n_{-1}-and-n_0 for s-family particles. Section 11 notes a pair of distinct rotational orders for n_{-2}-n_{-1}-and-n_0 for some q-family particles. For example, items related to item (11.29) establish an n_{-2}-n_{-1}-and-n_0 handedness/chirality for quarks. Up and down exhibit one handedness/chirality. Anti-down and anti-up exhibit the other handedness/chirality. Section 13 brings together the 2 uses of the rotational order. [for example, items following item (13.48)]

We discuss possibilities that reactions can convert quarks to leptons and vice-versa

Some \$o(−2) and \$o(−1) basic particles have charges with magnitudes specified by |Q'| = 1/3, 2/3, or 4/3. If all we consider is conservation of charge, any such o-family particle could intermediate reactions that convert quarks to leptons or leptons to quarks. Such a transformation correlates with changing the sign or Ω for a fermion. We do not discuss further herein the extent to which such transformations occur.

We suggest research

SOR.13.1 Detect or rule out (to some confidence level) that \$o(−2) and \$o(−1) bosons can covert a q-family S=1/2 fermion from anti-quark to quark and vice-versa.

SOR.13.2 Rule out (to some confidence level) or detect that \$o(−2) and \$o(−1) bosons can covert a q-family S=1/2 fermion (quark) to an l-family fermion (lepton) and vice-versa.

SOR.13.3 Determine the extents to which w-, h-, and o-family-mediated interactions (and/or e-family-coherence-mediated interactions) account for observed P violations, CP violations, or other such violations.

STR.13.1 To what extent do \$o%-mediated interactions correlate in strength with measured CP and/or P violations?

STR.13.2 To what extent do some o-family-mediated interactions correlate in strength with CP and/or P violations for which the Standard Model can account?

STR.13.3 To what extent might people benefit by considering that an interaction of a fermion with a 4w1 (or a 4w2, or 4w3) basic boson erases - from the fermion - 1 generation (or charge) and paints - on to the fermion - 1 generation (or charge)?

STR.13.4 To what extent might people benefit by considering that an interaction of a fermion with a \$o(0) (or a \$o(−2) or a \$o(−1)) basic boson erases - from the fermion - 1 generation (or charge) and paints - on to the fermion - 1 generation (or charge)?

STR.13.5 To what extent might people benefit by considering that an interaction of a fermion with a \$o(−4) or a \$o(−3) basic boson erases - from the fermion - 1 unit of property-3 and paints - on to the fermion - 1 unit of property-3?

STR.13.6 To what extent might people benefit by considering that an interaction of a fermion with a \$o(−6) or a \$o(−5) basic boson erases - from the fermion - 1 unit of magnetic moment and paints - on to the fermion - 1 unit of magnetic moment?

STR.13.7 To what extent might people benefit by considering that an interaction of a fermion with a \$o(−8) or a \$o(−7) basic boson erases - from the fermion - 1 unit of property-1 and paints - on to the fermion - 1 unit of property-1?

Section 14 Kinematics of some bosons

Abs.14.1 IOM correlates with kinematics of e-family and s-family bosons.

Context

We characterize, with respect to kinematics, some work above

Above, we do not much address details of quantum kinematics of particles or other objects.

We anticipate opportunities, with respect to kinematics, to extend work above

Here, we explore IOM that correlate with quantum kinematics. We correlate results with e- and s-family bosons.

Core

We focus on both time-correlated and space-correlated oscillators

We explore IOM(−2;−2,3;3). We use a hybrid QM-type-CS and QM-type-CL approach.

The next items match a time-correlated set of oscillators with a space-correlated set of oscillators. Here, r is a radial space-like coordinate and η has dimensions of length. Here, t' is a radial time-like coordinate and ï has dimensions of time.

$$\xi\,\Psi(r,t') = (\xi_0/2)\,(\,-\eta^2\,\nabla_{r}{}^2 + \eta^{-2}r^2\,)\,\Psi(r,t') \quad (14.1)$$
$$\upsilon\,\Psi(r,t') = (\upsilon_0/2)\,(\,-\ddot{\iota}^2\,\nabla_{t'}{}^2 + \ddot{\iota}^{-2}t'^2\,)\,\Psi(r,t') \quad (14.2)$$
$$\upsilon\,\Psi(r,t') = \xi\,\Psi(r,t') \quad (14.3)$$
$$D_e = 3 = D_p \quad (14.4)$$

For D_p=3, the next items provide an alternative form of the Laplacian operator. Here, r_2 provides a radial coordinate in 2 dimensions, φ provides the related angular coordinate, and x provides a linear coordinate for the third dimension. And, we use coordinates y and z for a linear description correlating with r_2^2. [Ref.14.1]

$$r_2^{-1}(\partial/\partial r_2)(r_2)(\partial/\partial r_2) + r_2^{-2}(\partial^2/\partial^2\varphi) + (\partial^2/\partial^2 x) \quad (14.5)$$
$$r_2^2 = y^2 + z^2 \quad (14.6)$$

For D_e=3, a similar result pertains. The next items show details.

$$t'^2 = t^2 + t_2^2 \quad (14.7)$$
$$t_2^2 = t_b^2 + t_c^2 \quad (14.8)$$
$$t_2^{-1}(\partial/\partial t_2)(t_2)(\partial/\partial t_2) + t_2^{-2}(\partial^2/\partial^2\varphi_{t2}) + (\partial^2/\partial^2 t) \quad (14.9)$$

The next items provide notation.

$$\nabla_{r2}{}^2 = r_2^{-1}(\partial/\partial r_2)(r_2)(\partial/\partial r_2) + r_2^{-2}(\partial^2/\partial^2\varphi) \quad (14.10)$$
$$\nabla_{t2}{}^2 = t_2^{-1}(\partial/\partial t_2)(t_2)(\partial/\partial t_2) + t_2^{-2}(\partial^2/\partial^2\varphi_{t2}) \quad (14.11)$$

The next item restates item (14.3). Here, we collect all operators featuring r or t on one side of the equal sign. Here, we collect all other operators on the other side of the equal sign.

$$(\upsilon_0/2)\,(\,-\ddot{\iota}^2(\partial^2/\partial t^2) + \ddot{\iota}^{-2}t^2\,) - (\xi_0/2)\,(\,-\eta^2(\partial^2/\partial x^2) + \eta^{-2}x^2\,) \quad (14.12)$$
$$=$$
$$(\xi_0/2)\,(\,-\eta^2(\nabla_{r2}{}^2) + \eta^{-2}r_2^2\,) - (\upsilon_0/2)\,(\,-\ddot{\iota}^2(\nabla_{t2}{}^2) + \ddot{\iota}^{-2}t_2^2\,)$$

The next item defines a symbol with dimensions of velocity.

$$v = \eta\,/\,\ddot{\iota} \quad (14.13)$$

The next items provide characteristics we posit for an edge solution. Here, δ denotes a Dirac delta function.

$$(\upsilon_0/2) = (\xi_0/2) \quad (14.14)$$
$$\Psi \propto \delta(x - vt) \quad (14.15)$$

The next item pertains.

$$(\upsilon_0/2)\,(\ddot{\iota}^{-2}t^2\,) - (\xi_0/2)\,(\eta^{-2}x^2\,) = 0 \quad (14.16)$$

The next items show relationships involving operators E and P corresponding to, respectively, energy and momentum for 1-dimensional motion.

$$E^2 \propto (\upsilon_0/2)\,(\,-\ddot{\imath}^2(\partial^2/\partial t^2)\,) \quad (14.17)$$
$$v^2P^2 \propto (\xi_0/2)\,(\,-\eta^2(\partial^2/\partial x^2)\,) \quad (14.18)$$
$$E^2 - v^2P^2 \propto (\upsilon_0/2)\,(\,-\ddot{\imath}^2(\partial^2/\partial t^2)\,)\, - (\xi_0/2)\,(\,-\eta^2(\partial^2/\partial x^2)\,) \quad (14.19)$$

The next items pertain to relevant solutions for the first of the last 2 terms in (14.12).

$$D_{p'} = 2 \quad (14.20)$$
$$\nu = -1/2 \text{ or } -1 \quad (14.21)$$

The next item pertains.

$$\Omega = \pm S(S + D_{p'} - 2) = \pm S^2 \quad (14.22)$$

The next items pertain to relevant solutions for the second of the last 2 terms in (14.12).

$$D_{e'} = 2 \quad (14.23)$$
$$\nu = -1/2 \text{ or } -1 \quad (14.24)$$

The next item pertains.

$$\Omega = \pm S(S + D_{e'} - 2) = \pm S^2 \quad (14.25)$$

We derive $E^2 - c^2P^2 = 0$ for photons and 4s% particles

For each of $D_{p'} = 2$ and $D_{e'} = 2$, the next items show the only edge solution. Here, $\nu = -1$.

S	Ω	D	D+2ν	
				(14.26)
1	1	2	0	(14.27)

We make interpretations based on the next item.

$$v = c \quad (14.28)$$

For these edge solutions, D+2ν=0. The next item pertains.

$$E^2 - c^2P^2 = 0 \quad (14.29)$$

The next items pertain.

The $D_{p'}$=2 edge solution (S=1, Ω=1 related to r_2) correlates with 4e2& (photons) (14.30)
The $D_{e'}$=2 edge solution (S=1, Ω=1 related to t_2) correlates with 4s% (the basis for gluons) (14.31)

We derive $E^2 - c^2P^2 = 0$ for the e-family

The next items pertain to expanding the focus from IOM(−2;−2,b_3=3;3) to IOM(−2;−2,b_3=9;9). We proceed mathematically via induction. We derive $E^2 - c^2P^2 = 0$ for the e-family.

We seek to show …	(14.32)
• For b_3 = 3, 5, 7, and 9, IOM describes $e%&, for $ = 5−(b_3−1)/2	(14.33)
Work above shows item (14.33) for b_3 = 3	(14.34)
• That work correlates with 4e2& (photons)	
Assume that we have established item (14.33) for some b_3, with 3 ≤ b_3 ≤ 7	(14.35)
To add 2 to b_3, we add to the after-the-equal-sign part of item (14.12) a new instance of item (14.10)	(14.36)
• The new instance features different 2-dimensional radial and angular coordinates than each of its previously considered siblings feature	
The new instance correlates with opening an oscillator pair	(14.37)
• The pair correlates with a new $D_{p'}$ =2 instance of item (14.27)	
The number of oscillators is appropriate for the $e%& that correlate with the value of $ = 5−(b_3−1)/2 for the new b_3 [Section 8]	(14.38)
We have shown, by induction, item (14.33)	(14.39)

Comments

We note some inside solutions

The next items show some inside solutions for ν=−1, for $D_{*ï}$ with ï = e or p. Here, 2S is an even integer.

S	Ω	D	D+2ν	(D+2ν) − 1	(14.40)
0	0	3	1	0	(14.41)
1	−1	4	2	1	(14.42)
2	−4	7	5	4	(14.43)
3	−9	12	10	9	(14.44)
4	−16	19	17	16	(14.45)
5	−25	28	26	25	(14.46)
6	−36	39	37	36	(14.47)
7	−49	52	50	49	(14.48)
8	−64	67	65	64	(14.49)

The next items show some inside solutions for ν=−1/2, for $D_{*ï}$ with ï = e or p. Here, 2S is an odd integer.

S	Ω	D	D+2ν	(14.50)
1/2	1/4	2	1	(14.51)
1/2	−1/4	3	2	(14.52)
3/2	−9/4	7	6	(14.53)
5/2	−25/4	15	14	(14.54)
7/2	−49/4	27	26	(14.55)
9/2	−81/4	43	42	(14.56)

We discuss the possibility of space-like (or faster-than-light-speed) behavior

The next item pertains.

Gss.14.1 In the sense of traditional perturbation models for interactions between elementary particles, people might consider that q-, s-, and o-family particles traverse space-like trajectories between vertices. (14.57)

We suggest research

STR.14.1 To what extent might people benefit by considering that $\Omega<0$-for-D_p aspects of IOM correlate with space-like (or faster-than-light-speed) behavior?

We list references

Ref.14.1 Wolfram Alpha, computational knowledge engine, Wolfram Alpha LLC, http://mathworld.wolfram.com/Laplacian.html.

Section 15 W-, h-, and o-family masses

Abs.15.1 IOM correlate with relative masses for w- and h-family bosons.
Abs.15.2 IOM may correlate with masses for o-family bosons.

Context

We do not know the extent to which traditional theory predicts masses for basic non-zero-mass bosons

People state masses for w- and h-family particles. People measure lower bounds for masses of leptoquarks. We do not know the extent to which traditional models predict such masses.

We anticipate developing approximate formulas relating masses of basic bosons

We explore the possibly that items following item (14.40) pertain to properties of w-, h-, and o-family bosons. We suggest a formula approximating masses of w- and h-family basic bosons. We suggest that extrapolation of the formula may correlate with o-family masses.

Core

We explore masses for w- and h-family basic bosons

The next items provide a formula that correlates with masses for w-family basic bosons (\$wχ, with \$=4) and the h-family basic boson (\$hχ, with \$=5 and χ=1).

$$m_Z = \text{the mass of a Z boson} \quad (15.2)$$
$$m^{``} = m_Z / 3 \quad (15.3)$$
$$(m(\$w\chi))^2 \approx (m^{``})^2 \times |f(\$\ddot{\imath}\chi)| \quad (15.4)$$
$$(m(\$h\chi))^2 \approx (m^{``})^2 \times |f(\$\ddot{\imath}\chi)| \quad (15.5)$$
$$f(4w1) = 9 \quad (15.6)$$
$$f(4w2) = f(4w3) \approx 7 \quad (15.7)$$
$$f(5w1) = 17 \quad (15.8)$$

The next items compare calculated and experimental masses. [Ref.15.1, Ref.15.2, and Ref.15.3]

Particle	Symbol	|f($ïχ)|	Calculated mass (GeV/c²)	Experimental mass (GeV/c²)	(15.9)
Z	4w1	9	91.188	91.1876±0.0021	(15.10)
W	4w2, 4w3	7	80.420	80.385±0.015	(15.11)
Higgs	5w1	17	125.325	125.3 ± 0.4 (stat) ± 0.5 (sys) [Ref.15.2] 126.0 ± 0.4 (stat) ± 0.4 (sys) [Ref.15.3]	(15.12)

We correlate w- and h-family masses with IOM math related to particle kinematics

Here (as we do for the e-family), we equate values of D+2ν with contributions to the squares of masses. Here, results correlate with |f($ïχ)|-column results that, respectively, items (15.10), (15.11), and (15.12) show.

The square of the mass of a Z boson correlates with the sum of (15.13)

- −1 = minus a $D_{e'}$=2 instance of item (14.41)
- +10 = plus a $D_{p'}$=2 instance of item (14.44)

The square of the mass of a W boson correlates with the sum of (15.14)

- −1 = minus a $D_{e'}$=2 instance of item (14.41)
- −2 = minus a $D_{e'}$=2 instance of item (14.42)
- +10 = plus a $D_{p'}$=2 instance of item (14.44)

The square of the mass of a Higgs boson correlates with the sum of (1 term) (15.15)

- +17 = plus a $D_{p'}$=2 instance of item (14.45)

To the extent these results pertain, we might attribute the item-(15.11) discrepancy regarding W-boson mass to use immediately above of no more than IOM(−8;−4,3;3). An extension to IOM(−8;−8,3;3) might bring in other considerations (for example, related to magnetic moment). The other considerations might reduce the calculated mass for W bosons.

Comments

We discuss possible o-family masses

We explore extending work above to correlate with o-family masses.
The next items provide features of 2 mutually exclusive possibilities.

O-family-1: The largest contribution to the magnitude of the mass of a 4o(0) boson is the S=5 item following item (14.40) (15.16)

- That item shows D+2ν = 26

O-family-2: The largest contribution to the magnitude of the mass of a 4o(0) boson is the S=3 item following item (14.40) (15.17)

- That item shows D+2ν = 10

The next items illustrate possible o-family-1 rules extending trends in items (15.13), (15.14), and (15.15). We consider aspects of this work to be speculative.

The square of the mass of a 4o(0) boson correlates with the sum of (15.18)

- +1 = plus a $D_{p'}$=2 instance of item (14.41)
- −26 = minus a $D_{e'}$=2 instance of item (14.46)

The square of the mass of a 4o(−2) boson (or a 4o(−1) boson) correlates with the sum of (15.19)

- +1 = plus a $D_{p'}$=2 instance of item (14.41)
- +2 = plus a $D_{p'}$=2 instance of item (14.42)
- −26 = minus a $D_{e'}$=2 instance of item (14.46)

The square of the mass of a 3o(0) boson correlates with the sum of (15.20)

- +1 = plus a $D_{p'}$=2 instance of item (14.41)
- +2 = plus a $D_{p'}$=2 instance of item (14.42)
- −37 = minus a $D_{e'}$=2 instance of item (14.47)

The square of the mass of a 2o(0) boson correlates with the sum of (15.21)

- +1 = plus a $D_{p'}$=2 instance of item (14.41)
- +2 = plus a $D_{p'}$=2 instance of item (14.42)
- +5 = plus a $D_{p'}$=2 instance of item (14.43)
- −50 = minus a $D_{e'}$=2 instance of item (14.48)

The square of the mass of a 1o(0) boson correlates with the sum of (15.22)

- +1 = plus a $D_{p'}$=2 instance of item (14.41)
- +2 = plus a $D_{p'}$=2 instance of item (14.42)
- +5 = plus a $D_{p'}$=2 instance of item (14.43)
- +10 = plus a $D_{p'}$=2 instance of item (14.44)
- −65 = minus a $D_{e'}$=2 instance of item (14.49)

The next items show, for the o-family-1 possibility, the magnitudes of some possible results, in units of GeV/c^2. The items show, for each particle shown, |f| (paralleling items following item (15.9)), a magnitude of mass (1×m), twice that magnitude (2×m), and thrice the magnitude (3×m).

Particle	O-family-1				
	|f|	1×m	2×m	3×m	(15.24)
4o(0)	25	152 GeV/c^2	304	456	(15.25)
4o(−2)	23	146	292	437	(15.26)
3o(0)	34	177	354	532	(15.27)
2o(0)	42	197	394	591	(15.28)
1o(0)	47	208	417	625	(15.29)

(15.23)

We discuss results for leptoquarks and o-family masses

We recall the notion that o-family particles must be created in groups of (at least) 2 or 3. We do not know the extent to which a binding energy might pertain to such a pair or triplet of particles.

Experimental results provide lower bounds for masses of various hypothesized leptoquarks. The next item notes a range of reported lower bounds (in GeV/c^2 and with minimum confidence levels of 95%). [Ref.15.4]

226 - 685 (15.30)

We do not know the extents to which leptoquarks would correlate with o-family single particles, o-family groups of co-created particles, and/or not correlate with the o-family.

We suggest research

SOR.15.1 Verify or rule out (to some confidence level) that much of the difference between the W-boson mass we calculate and the observed W-boson mass correlates with a non-zero magnetic moment for W bosons.
STR.15.1 Estimate a value for the magnetic moment of W bosons.
STR.15.2 Develop a model specifying values of D+2ν correlating with o-family masses.
STR.15.3 To what extent might people benefit from considering that an antiparticle for each of 4w1, 5h1, and the various $o(0) particles (4≥$≥1) has an Ω that is the negative of the Ω for the particle?

We list references

Ref.15.1 J. Beringer et al. (Particle Data Group), *PR D86*, 010001 (2012) and 2013 partial update for the 2014 edition (URL: http://pdg.lbl.gov). (http://pdg.lbl.gov/2013/tables/rpp2013-sum-gauge-higgs-bosons.pdf)
Ref.15.2 CMS collaboration (2012). "Observation of a new boson at a mass of 125 GeV with the CMS experiment at the LHC". *Physics Letters B* 716 (1): 30–61. arXiv:1207.7235. Bibcode:2012PhLB..716...30C. doi:10.1016/j.physletb.2012.08.021.
Ref.15.3 ATLAS collaboration (2012). "Observation of a New Particle in the Search for the Standard Model Higgs Boson with the ATLAS Detector at the LHC". *Physics Letters B* 716 (1): 1–29. arXiv:1207.7214. Bibcode:2012PhLB..716....1A. doi:10.1016/j.physletb.2012.08.020.
Ref.15.4 J. Beringer et al. (Particle Data Group), *PR D86*, 010001 (2012) and 2013 partial update for the 2014 edition (URL: http://pdg.lbl.gov).

Section 16 Q- and l-family masses

Abs.16.1 A formula approximates masses of quarks and charged leptons.
Abs.16.2 The formula for masses of charged basic fermions may provide masses for neutrinos.
Abs.16.3 IOM may approximately correlate with relative masses for quarks and charged leptons.

Context

We do not know the extent to which traditional theory predicts masses for basic fermions

Regarding the l-family, people discuss a formula [item (16.61)] linking the masses of charged leptons. People do not know masses for neutrinos. People estimate upper bounds on neutrino masses.

Regarding the q-family, people measure approximate masses for quarks.

We do not know of traditional theory that correlates with relative masses for l- and q-family particles.

We anticipate developing approximate formulas correlating masses of basic S=1/2 fermions

We array l- and q-family masses in a way that suggests a pattern. We develop a formula that correlates with the pattern. Based on the possibility that the formula extends to neutrinos and on upper bounds for neutrino masses, we suggest possible masses for neutrinos.

Core

We show formulas that correlate with S=1/2-q-family and charged-l-family masses

Edge solutions correlate with q- and l-family particles. Above, we discuss the possibilities that considering ψ(r), with r<0, makes sense regarding such particles. [(4.48)] We note possibilities for roles for exponential and trigonometric functions.

The next item defines notation regarding items following item (11.19).

m(M",M') denotes a calculated mass correlating with a position in a table that includes positions for S=1/2 charged basic fermions (16.1)

The next item pertains to trends in items following item (11.19).

Gss.16.1 For charged leptons (either M'=−3 or M'=+3), people can benefit by correlating the range −1≤M"≤3 with an L=2 system. (16.2)

The next item follows. For the L=2 system, −2≤M"−1≤+2. We seem to find that no solution with d(0)=−d(2) appeals. We think such a solution would correspond to a trigonometric function that is anti-symmetric with respect to M"−1. The next item features an expression that can correspond to a symmetric trigonometric function.

Gss.16.2 For the L=2 system that includes charged leptons, m(M",−3) ∝ $e^{M''\zeta''}(1+d(M''))$, in which −1≤M"≤3, d(0)=d(2), and d(−1)=d(1)=d(3)=0. (16.3)

The next items show results for M'=−3 charged leptons and related values of M". Here, we use the experimental masses for the electron and muon to calculate ζ". Then, we calculate m\`. (To do so, we use an experimentally acceptable calculated mass for the tauon. We use items (16.6) and (16.7) to calculate a tauon mass.) Then, we calculate d(0).

$$m(M'',-3) = m` \times \exp((M''+1)\zeta'') \times (1+d(M'')) \quad (16.4)$$
$$\zeta'' = (1/2)\log(m_{muon}/m_e) \approx 2.665799 \quad (16.5)$$
$$m_{tauon} \approx \beta\, m_e \quad (16.6)$$
$$\beta \approx 3.47714\times10^3 \quad (16.7)$$
$$m` = m_{tauon} / \exp(4\zeta'') \approx 4.155987\times10^{-2}\ \mathrm{MeV}/c^2 \quad (16.8)$$
$$1 + d(0) = m_e / (m` \exp(\zeta'')) \quad (16.9)$$
$$d(2) = d(0) \approx -0.144926 \quad (16.10)$$
$$d(-1) = d(1) = d(3) = 0 \quad (16.11)$$

The next items correlate with a possible role for trigonometric functions.

$$d(M'') \approx d'' \times (1/2) \times (\cos(M''\pi) + 1) \quad (16.12)$$
$$d'' = d(2) \approx -0.144926 \quad (16.13)$$

The next item provides a calculated number not directly correlated with a particle.

$$m(1,-3) \approx 8.59326\ \mathrm{MeV}/c^2 \quad (16.14)$$

The next items extend this work to include masses for quarks and antiparticles. Here, α denotes the fine-structure constant.

$$m(M'', M') \approx m` \times \exp((M''+1)\zeta'') \times (1+d(M'')) \times \exp((1/4)\log(1/\alpha)(1+M'')(3+M')) \times (1+d(M'',M')), \quad \text{for } M'\leq 0 \tag{16.15}$$

$$m(M'', M') = m(M'', -M') \tag{16.16}$$

The next items finish defining the formula correlating with approximate masses for quarks and charged leptons. Here, we use item (11.27). That is, for any relevant M", quarks and charged leptons correlate with an L=3 system in which −3≤M'≤3. Here, we use item (16.14). We specify d(M",M') adequate to fit known experimental results. Item (16.20) correlates with antiparticles and particles having identical masses.

$$d(M'',M') \tag{16.17}$$

$$\approx (-\sin(2\pi M'/3)/(\sin(2\pi/3)) \times (-\cos(\pi M'')) \times (d')\exp(-(M''+1)\log(\zeta'))$$

$$\approx (\sin(2\pi M'/3)/(\sin(2\pi/3)) \times (\cos(\pi M'')) \times (d')\exp(-(M''+1)\log(\zeta')) \quad \text{for } -3\leq M'\leq 0$$

$$d' \approx 0.5 \tag{16.18}$$

$$\zeta' \approx 0.04 \tag{16.19}$$

$$d(M'',M') = d(M'',-M') \tag{16.20}$$

The next items pertain.

$$-d(0,-1) = d(0,-2) \sim 0.2 \tag{16.21}$$

$$-d(1,-1) = d(1,-2) \sim -0.08 \tag{16.22}$$

$$-d(2,-1) = d(2,-2) \sim 0.032 \tag{16.23}$$

$$d(M'',M') = 0, \text{ otherwise within } 0\leq M''\leq 2, -3\leq M'\leq 0 \tag{16.24}$$

The next 2 sets of items compare calculated numbers with experimental numbers. [Ref.6.5 and Ref.16.1] The items show masses in units of MeV/c^2. For each particle, the bottom number (calc) comes from our calculations. Except regarding M"=2 quarks, the upper number (exp) comes from experiments. For M"=2 quarks, Ref.16.1 provides two possible ranges for quark masses. The upper range is based on mass-independent subtraction scheme(s) (MS). For M"=2, M'=−2, the first mass is the MS running mass and the second mass is the 1S mass. For M"=2, M'=−1, the first mass is labeled MS from cross-section measurements and the second mass is from direct measurements.

		M'	−3	...	...	
M"						(16.26)
0	exp		0.510998928±0.000000011			(16.27)
0	calc		0.510998928 MeV/c^2			(16.28)
1	exp					(16.29)
1	calc					(16.30)
2	exp		105.6583715±0.0000035			(16.31)
2	calc		105.6583715			(16.32)
3	exp		1776.82±0.16			(16.33)
3	calc		1776.81			(16.34)

(16.25) is the M' header row.

		M' ...	−2	−1	(16.35)
M"					(16.36)
0	exp		$2.3^{+0.7}_{-0.5}$	$4.8^{+0.7}_{-0.3}$	(16.37)
0	calc		2.10 MeV/c^2	4.79	(16.38)
1	exp		95±5	$(1.275\pm0.025)\times10^3$	(16.39)
1	calc		92.5	1.272×10^3	(16.40)
2	exp		$(4.18\pm0.03)\times10^3$	$(160^{+5}_{-4})\times10^3$	(16.41)
			$(4.65\pm0.03)\times10^3$	$(173.5\pm0.6\pm0.8)\times10^3$	
2	calc		4.367×10^3	164.1×10^3	(16.42)
3	exp				(16.43)
3	calc				(16.44)

Possibly, m(M",M') correlates with each of the 9 masses for quarks and charged leptons. The next item lists relevant numbers that may seem not to be widely used in mathematics.

m`, ζ", d", α, d', and ζ' (16.45)

We discuss possible masses for neutrinos

The next items pertain.

Gss.16.3	The up-to-3 neutrino masses correlate with up to 3 numbers of the form m(M",0), in which ύ−3≤M"≤ύ, for which M"=ύ correlates with the largest mass not ruled out by observations.	(16.46)
Gss.16.4	The formula for m(M", M') has meaning for M"<−1. The trigonometric-like pattern for d(M") continues throughout the range −6≤M"≤3. d(M",0) = 0.	(16.47)

The next items provide estimated values of masses of neutrinos. The items show masses in units of eV/c^2. People interpret observations as implying that no more than 3 neutrinos (plus 3 possible anti-neutrinos) exist. [Ref.16.2] Math above regarding generations correlates with exactly 3 neutrinos (plus, if applicable, 3 antiparticles). People interpret observations as implying that the sum of the up-to-3 neutrino masses does not exceed 0.28 eV/c^2. [Ref.16.3 and Ref.16.4] No more that 3 of the masses in the next items pertain. We do not know of relevant experimental results.

		M'	0	(16.48)
M"				(16.49)
−6	exp			(16.50)
−6	calc		6×10^{-10} eV/c^2	(16.51)
−5	exp			(16.52)
−5	calc		4×10^{-7}	(16.53)
−4	exp			(16.54)
−4	calc		2×10^{-4}	(16.55)
−3	exp			(16.56)
−3	calc		1×10^{-1}	(16.57)

Comments

We point to an involvement of square roots of masses

The next items restate items that include and follow item (16.4).

$m(M'', M') \approx m` \times (m_{muon} / m_e)^{(1/2)(M''+1)} \times (1+d(M'')) \times (1/\alpha)^{(1/4)(1+M'')(3+M')} \times (1+d(M'',M'))$ (16.58)

- $d(M'') \approx d'' \times (1/2) \times (\cos(M''\pi) + 1)$
- $d(M'',M') \approx (\sin(2\pi M'/3)/(\sin(2\pi/3)) \times (\cos(\pi M'')) \times (d')\exp(-(M''+1)\log(\zeta'))$
- for M'<0; plus, $m(M'', M') = m(M'', -M')$
- with ... (16.59)
    - $m` \approx 4.155987\times10^{-2}$ MeV/c^2
    - $d'' \approx -0.144926$
    - $\alpha \approx 7.29735\times10^{-3}$ (the fine-structure constant)
    - $d' \approx 0.5$
    - $\zeta' \approx 0.04$

For charged leptons, the next item pertains. The formula involves integer powers of the square roots of 2 masses.

$$m(M'', M') \approx m` \times ((m_{muon} / m_e)^{(1/2)})^{(M''+1)} \times (1+d(M'')) \quad (16.60)$$

The next item shows the Koide formula. People correlate this formula with aspects of the Standard Model.

$$(m_e + m_{muon} + m_{tauon}) / (m_e^{1/2} + m_{muon}^{1/2} + m_{tauon}^{1/2})^2 \approx 2/3 \quad (16.61)$$

The next item reflects nominal numbers we use, including the calculated value for m_{tauon}. The uncertainty-range may not be accurate.

$$(m_e + m_{muon} + m_{tauon}) / (m_e^{1/2} + m_{muon}^{1/2} + m_{tauon}^{1/2})^2 \approx 0.6666579(2) \quad (16.62)$$

We discuss possible correlations between powers of β and interaction strengths

The next items correlate with items (6.3) and (6.6). In particular, in item (16.65), γ=5 for tauons.

Force	Channels	Electron relative vertex strength per channel	Electron γ for $\beta^{-\gamma}$	Tauon relative vertex strength per channel	Tauon γ for $\beta^{-\gamma}$	(16.63)
4e2&	4	$1 = \beta^{-0}$	0	$1 = \beta^{-0}$	0	(16.64)
3e4&	3	β^{-6}	6	β^{-5}	5	(16.65)

The next items include information about the muon.

Particle	Symbol for mass m(M",M')	mass/m_e	M"	(16.66)
electron	m(0,−3) = m_e	$\beta^{0/3}$	0	(16.67)
-			1	(16.68)
muon	m(2,−3)	$\sim\beta^{(2/3)\cdot(1-0.02)}$ or $\sim 0.9\cdot\beta^{2/3}$	2	(16.69)
tauon	m(3,−3)	$\beta^{3/3}$	3	(16.70)

We explore an alternative approximation for masses of charged fermions

We explore another approach to modeling masses for charged basic fermions.

Various D+2ν for boson fields correlate with w- and h-family masses. [Section 15] Section 14 lists D for fermion fields. For ν=−1/2, the smallest values of D+2ν are 1, 2, 6, 14 and 26. [items following item (14.50)]

The next items sketch the model. We model relationships between squares of masses. For the m(M",−3), we use numbers calculated above. [items following item (16.25)]

$$(m / m(M'',M'))^2 \approx f(\acute{\upsilon}(M''),n(M'',M')) / f(\acute{\upsilon}(M''),n(M'',-3)) \quad (16.71)$$

$$f(\acute{\upsilon},n) = e^{\acute{\upsilon}|n|} \quad (16.72)$$

$$\acute{\upsilon}(0) \approx 0.9114 \,,\, \acute{\upsilon}(1) \approx 0.911 \,,\, \acute{\upsilon}(0) \approx 0.922 \quad (16.73)$$

The next items show values of n based on guesses as to which values of D+2ν pertain.

	M'=−3	M'=−2	M'=−1	(16.74)
M"=0	1 = 1	−4 = 2−6	−6 = −6	(16.75)
M"=1	2 = 2	−7 = 1+6−14	−13 = 1−14	(16.76)
M"=2	3 = 1+2	−11 = 1+14−26	−19 = 1+6−26	(16.77)

The next items show calculated masses (in MeV/c^2). (The calculated masses for electrons and muons are the same as shown above. The M"=1, M'=−3 value comes from item (16.14).) Only the mass for the strange quark (M"=1, M'=−2) seems out of experimental uncertainty-range. That calculated mass differs by about 9% (9 percent) from the mass calculated above. Each other number shown here differs by less than 5% (5 percent) from the respective calculated mass shown above.

	M'=−3	M'=−2	M'=−1	(16.78)
M"=0	0.5109989...	2.00	4.99	(16.79)
M"=1	8.59 MeV/c^2	8.38×10^1	1.29×10^3	(16.80)
M"=2	$(1.05658...)\times10^2$	4.22×10^3	1.69×10^5	(16.81)

We suggest research

SOR.16.1 Measure neutrino masses.

SOR.16.2 Rule out (to some confidence level) or detect the existence of M"≥−2, M'=0 particles.

STR.16.1 What would be the impact of detection of M'=0 baryonic-matter fermions having M"≥−2?

STR.16.2 To what extent do masses of q- and l-family fermions correlate with masses of w- and h-family bosons?

STR.16.3 To what extent does the appearance, in a formula for the ratios of masses of charged leptons, of a power of the ratio of the square roots of 2 lepton masses correlate with the possible applicability of the Koide formula? [(16.58) and (16.61)]

STR.16.4 Develop a model specifying values of D+2ν, ύ(M"), and adjustments to use for calculating quark masses. [items including and following item (16.71)]
STR.16.5 Extend work regarding STR.4.1 to model S=1/2 basic fermion masses.
STR.16.6 Develop models people can use to estimate masses for S≥3/2 basic fermions.

We list references

Ref.16.1 J. Beringer et al. (Particle Data Group), *Phys. Rev. D86*, 010001 (2012). (http://pdg.lbl.gov/2012/tables/rpp2012-sum-quarks.pdf)
Ref.16.2 J. Beringer et al. (Particle Data Group), *PR D86*, 010001 (2012) and 2013 partial update for the 2014 edition (URL: http://pdg.lbl.gov). (http://pdg.lbl.gov/2013/tables/rpp2013-sum-leptons.pdf)
Ref.16.3 S. Thomas, F. Abdalla, and O. Lahav, Upper Bound of 0.28 eV on the Neutrino Masses from the Largest Photometric Redshift Survey, *Phys. Rev. Lett. 105*, 031301, 2010. (http://arxiv.org/abs/0911.5291)
Ref.16.4 A. Melchiorri, Constraints on Neutrino Physics from Planck, European Space Agency, http://www.rssd.esa.int/SA/PLANCK/docs/eslab47/Session06_CMB_Cosmology_and_Fundamental_Physics/47ESLAB_April_04_17_30_Melchiorri.pdf.

Section 17 E-family interaction strengths

Abs.17.1 Formulas provide approximate relative strengths for interactions mediated by e-family basic bosons and some e-family coherences.

Context

We note the extent to which we describe e-family interactions above

Above, we describe relative interaction strengths for the first 2 members of the photon-graviton series of 4 basic bosons. The work discusses interactions between 2 electrons.

We anticipate extrapolating to other relative interaction strengths

Here, we indicate approximate interaction strengths for interactions carried by e-family members other than just 4e2& and 3e4&.

Core

We discuss strengths of photon-graviton series basic bosons

The next item pertains. The M"=0 leptons are electrons and positrons.

Gss.17.1 For photon-graviton series basic bosons, for interactions between 2 M"=0 leptons, the relative vertex strength per channel follows a pattern established by the relative vertex strengths per channel for photons and gravitons. (17.1)

The next items show the pattern. For $e%& with 3≥$, each relative vertex strength per channel equals β^{-6} times the preceding one.

Interaction	Channels	Relative vertex strength per channel	Relative interaction strength per channel	(17.2)
4e2&	4	$1 = \beta^{-0}$	$1 = (\beta^{-0})^2$	(17.3)
3e4&	3	β^{-6}	$(\beta^{-6})^2$	(17.4)
2e6&	2	β^{-12}	$(\beta^{-12})^2$	(17.5)
1e8&	1	β^{-18}	$(\beta^{-18})^2$	(17.6)

The next item pertains.

Gss.17.2 Vertex strengths scale per particle properties. (17.7)

The next items pertain. These items combine item (6.25) and following items, item (17.2) and following items, and item (17.7).

Interaction	Channels	Object property	Symbol for object property	Relative vertex strength per channel	(17.8)
4e2&	4	Charge	Q	$\beta^{-0}\, Q / \lvert q_e \rvert$	(17.9)
3e4&	3	Property-3	m	$\beta^{-6}\, m / m_e$	(17.10)
2e6&	2	Magnetic moment	m_{mm}	$\beta^{-12}\, m_{mm} / m_{mm,e}$	(17.11)
1e8&	1	Property-1	propterty-1	β^{-18} |(property-1)| / |(property-1)$_e$|	(17.12)

with ... (17.13)

- $m_{mm,e} = g_s\hbar/2 = \hbar$
- $|(\text{property-1})_e| = 1$

We discuss strengths for e-family members other than photon-graviton series basic bosons

The next items pertain.

$F_ύ(R)$ denotes the force associated with ύe(2·(5-ύ))& between 2 electrons separated by a distance R (17.14)

For example, for ύ=4, (17.15)

- The interaction is carried by 4e2&
- $F_ύ(R) = (q_e^2/4\pi\varepsilon_0)\,(1/R)^2$

The next items show a ratio of magnitudes of forces pertaining to interactions between 2 electrons. Here, λ_{*3} is any positive length.

$$F_4(R)\ \ F_3(R) \,/\, F_3(R) = F_4(R) \quad (17.16)$$

$$F_4(R)\ \ F_3(R) \,/\, F_3(R) = F_4(\lambda_{*3})\ (\lambda_{*3}/R)^2 \quad (17.17)$$

The left side of (17.17) is proportional to the ratio of the magnitude of the 3e24& interaction between 2 electrons to the magnitude of the 3e4& interaction between 2 electrons.

The next items show similar expressions. The first item pertains to 2e246& and 2e6&. The second item pertains to 1e2468& and 1e8&.

$$F_4(R)\ F_3(R)\ F_2(R)\ /\ F_2(R) = F_4(\lambda_{*2})\ F_3(\lambda_{*2})\ (\lambda_{*2}/R)^4 \quad (17.18)$$
$$F_4(R)\ F_3(R)\ F_2(R)\ F_1(R)\ /\ F_1(R) = F_4(\lambda_{*1})\ F_3(\lambda_{*1})\ F_2(\lambda_{*1})\ (\lambda_{*1}/R)^6 \quad (17.19)$$

We discuss lengths at which photon-graviton series forces match maximal-% series forces

The next items pertain.

Gss.17.3	For interactions between 2 electrons, the strengths of 3e4& and 3e24& are roughly equal at a particle separation of λ_3.	(17.20)
Gss.17.4	For interactions between 2 electrons, the strengths of 2e6& and 2e246& are roughly equal at a particle separation of λ_2.	(17.21)
Gss.17.5	For interactions between 2 electrons, the strengths of 1e8& and 1e2468& are roughly equal at a particle separation of λ_1.	(17.22)

We discuss magnitudes of maximal-% series forces

The next items pertain to interactions between 2 electrons.

$$4e2\&(R) = (q_e^2/4\pi\varepsilon_0)\ (1/R)^2 \quad (17.23)$$
$$3e24\&(R) = 3e4\&(R)\ (\lambda_3/R)^2 = 4e2\&(R)\ ((4/3)\beta^{12})^{-1}\ (\lambda_3/R)^2 \quad (17.24)$$
$$2e246\&(R) = 2e6\&(R)\ (\lambda_2/R)^4 = 4e2\&(R)\ ((4/2)\beta^{24})^{-1}\ (\lambda_2/R)^4 \quad (17.25)$$
$$1e2468\&(R) = 1e8\&(R)\ (\lambda_1/R)^6 = 4e2\&(R)\ ((4/1)\beta^{36})^{-1}\ (\lambda_1/R)^6 \quad (17.26)$$

The next items extend items starting with item (17.8). Here, for each property ύ, ύ(1) denotes that property for object 1. Here, for each property ύ, ύ(2) denotes that property for object 2. These items pertain to the magnitudes of the various $e%&. For 4e2, ± = 1. For other $e%& for which % includes a 2, we are uncertain regarding signs.

If % contains the following symbol ...	... multiply the 2-electron result by ...	(17.27)
2	$\pm\ Q(1)\ \ Q(2)\ /\ (q_e)^2$	(17.28)
4	(property-3)(1) (property-3(2) / $(m_e)^2$	(17.29)
6	$m_{mm}(1)\ \ m_{mm}(2)\ /\ (m_{mm,e})^2$	(17.30)
8	(property-1)(1) (property-1)(2) / ((property-1)$_e$)^2	(17.31)

Comments

We discuss phenomena related to coherences

For a pair of correlated photons created by the annihilation of an electron-positron pair, an object can interact with 1 of the photons and leave the other photon alone. [Section 18 provides an example.] For a laser-created photon state with at least 2 excitations, an object can absorb 1 unit of excitation and leave the other units alone. We interpret discussion [items including and following item (13.29)] related to physics some people correlate with axions to correlate with the notion that an object can interact with the 3e4 mode within a 3e24 in such a way as to leave the 4e2 alone.

Instances of effects of coherences involve compound (or even seemingly unrelated) objects. Instances of coherences arise from addition of a unit of an e-family mode to an already existing coherence or e-family basic boson. Instances of coherences change based on annihilation of a single excitation of a single mode.

We discuss magnitudes of strengths of some e-family interactions

The next items provide numbers for relative vertex strengths and relative interaction strengths for interactions between 2 electrons, 2 positrons, or 1 electron and 1 positron. The second column notes logarithms of strengths relative to the strength per channel of a vertex for electromagnetism. The third column notes logarithms of strengths relative to the strength per channel of a vertex for gravity. The fifth column notes logarithms of interaction strengths relative to the strength of electromagnetism. The rightmost column notes logarithms of interaction strengths relative to the strength of gravity.

Interaction	Log_{10} (relative vertex strength per channel)	Log_{10} (relative vertex strength per channel)	Channels	Log_{10} (relative interaction strength)	Log_{10} (relative interaction strength)	(17.32)
4e2&	0.0	21.2	4	0.0	+42.6	(17.33)
3e4&	−21.2	0.0	3	−42.6	0.0	(17.34)
2e6&	−42.5	−21.2	2	−85.3	−42.7	(17.35)
1e8&	−63.7	−42.5	1	−128.1	−85.5	(17.36)

For each of the columns noting relative vertex strengths per channel, successive numbers differ by approximately −21.2474. The next item pertains.

$$Z\cdot 10^{-21.2474\ldots} \approx 6.7584 \qquad (17.37)$$

The next items show results from using items (17.20), (17.21), and (17.22). Relative to the strength of gravity (3e4&), entries show the log-base-10 of approximate values for the strengths of interactions between 2 electrons that are 1 m apart. Values ignore signs that would connote attraction or repulsion.

E-family member	Log_{10} (magnitude of relative interaction strength), for 2 electrons, 1 m apart	(17.38)
4e2&	+42.6	(17.39)
3e4&	0.0	(17.40)
3e24&	−113.7	(17.41)
2e246&	−358.5	(17.42)
1e2468&	−691.6	(17.43)

The next items show values pertaining to hypothetical \$e. Each \$e reflects the concept of extrapolating by removing elements from % in a series for which \$e% has non-zero numbers of elements in %. For Ø denoting the null set (or empty set), conceptually, \$e denotes \$eØ. Presumably, the \$e values depend neither on the properties of 2 interacting electrons nor on the distance between the 2 interacting electrons. Values ignore signs that would connote attraction or repulsion. The value for 3e4& pertains to 2 electrons separated by 1 m.

E-family member	$\log_{10}$(relative interaction strength)	
		(17.44)
3e4&	0.0	(17.45)
4e	112.2	(17.46)
3e	113.7	(17.47)
2e	115.2	(17.48)
1e	116.6	(17.49)

The next item pertains to items (17.46) and (17.47).

$$10^{113.7} / 10^{112.2} \approx 34.26 \approx (Z')^2 \qquad (17.50)$$

For $e results, the numbers do not depend on the choice of 2 interacting small objects or on the distances between the 2 objects.

We interpret the near equality of the $e numbers as not contradicting items (17.20), (17.21), and (17.22). Perhaps, better choices of lengths at which electron-electron forces have similar strengths (better than $\lambda_{\acute{\upsilon}}$, for $\acute{\upsilon}$ = 3, 2, and 1) would lead both to more nearly equal values in items (17.46), (17.47), (17.48), and (17.49) and to better correlations with nature.

We discuss a possible example of a perturbation technique

We explore a possibility that a perturbation-like technique applies. In particular, we note a possible approximation for α, the fine-structure constant.

The next items define the approximation.

$$\alpha \approx \Sigma \text{ (channel ratio) } \kappa^{\gamma} (2\pi)^{\gamma'} \qquad (17.51)$$

A channel ratio denotes the ratio of channels for $ (in $e%&) and $=4 (in 4e2&) (17.52)

$$\kappa = 2 \qquad (17.53)$$

Here, we choose γ for κ^{γ} to match series that feature powers of β. We choose κ = 2 to correlate with the opening (term by term) of oscillator pairs. We choose γ' to obtain an approximate result. The next items show numbers.

Channel ratio	γ in κ^{γ}	γ' in $(2\pi)^{\gamma'}$	Single term	Σ = cumulative sum of terms	(Σ − α) / α	(17.54)
3/4	−12	2	7.22871×10^{-3}	7.2287×10^{-3}	-9.4059×10^{-3}	(17.55)
2/4	−24	4	4.64483×10^{-5}	7.2752×10^{-3}	-3.0408×10^{-3}	(17.56)
1/4	−36	8	8.83688×10^{-6}	7.2840×10^{-3}	-1.8299×10^{-3}	(17.57)

We suggest research

STR.17.1 Develop theory sufficient to predict choices - attraction, repulsion, or neither - for each e-family interaction between 2 particles.

STR.17.2 To what extent might people benefit by exploring theories in which the property of charge correlates with concepts of curvature for a space time in which the number of spatial dimensions is 1?

STR.17.3 To what extent might people benefit by considering that effects of $e%& interactions might be modeled in a flat space time for which there are no more than 3 time-like dimensions and no more than 2(5−$)+1 space-like dimensions? (Here, the term flat refers to a Minkowski-like metric with each off-diagonal term having a value of 0. Here, one can consider 4 cases - $ = 4, 3, 2, and 1.)

STR.17.4 To what extent should people associate phenomena related to $e%& with people's notions of space-time froth? (For example, do 4e2&-related phenomena provide for supposed froth on a scale people associate with the Planck length?)

STR.17.5 To what extent might people use parallels to the electromagnetic vector potential when describing gravity and other $e%& interactions? (For example, for 3e4&, property-3 could be an analog to charge; and, a concept of a property-3 current could lead to a property-3-based analog to magnetic fields.)

STR.17.6 To what extent might people extend the Standard Model to include gravity and non-traditional e-family interactions? (For example, can people base such an extension on potentials, currents, and so forth suggested by STR.17.5?)

STR.17.7 Develop a suitable IOM perturbation theory (possibly based on something like Feynman diagrams) for e-family interactions.

STR.17.8 Use such an IOM perturbation theory [STR.17.7] to estimate magnetic-moment anomalies. [items (6.47) and (6.48)]

Section 18 Examples of interactions

Abs.18.1 We illustrate interactions involved in fermion-anti-fermion annihilation.
Abs.18.2 We illustrate interactions involved in neutrino oscillations.
Abs.18.3 We illustrate mechanics of channels.

Context

We discuss in this paper various interactions

Above, we allude to various interaction vertices and various types of interactions.

We illustrate various types of interactions

Here, we show examples of interaction vertices and types of interactions.

Core

We discuss a weak-interaction vertex

The next items pertain to a vertex in which an electron-neutrino and a W^- produce an electron. The description does not depend on choosing an n-type for the neutrino.

- An electron-neutrino enters (18.1)
- A W^- enters

| ; n_0=2 , n_1=# , n_2=1 , n_3=# ; >

- An electron exits (18.2)
- A w-family ground state remains

| ; n_0=1 , n_1=0 , n_2=0 , n_3=0 ; >

We discuss electron-positron annihilation

The dominant mode for electron-positron annihilation produces 2 photons. In a traditional Feynman diagram, an electron enters, 2 photons exit, and the positron exits as a continuation of the electron. People may say that the positron goes backward in time.

In an IOM treatment, the next items pertain.

- An electron enters (18.3)
- A positron enters
- The l-family field absorbs each of the 2 charged leptons (18.4)
- 2 correlated photons are produced (18.5)
- The 2 correlated photons exit

The next items describe a scenario underlying the annihilation reaction.

Four W bosons are created in the form (18.6)

| ; n_0=3 , n_1=# , n_2=2 , n_3=# ; > plus | ; n_0=3 , n_1=# , n_2=# , n_3=2 ; >

Each of the electron and positron absorbs a W boson, resulting in the following (18.7)

- The l-family field gains 2 in fermion count
- The w-family state becomes

| ; n_0=2 , n_1=# , n_2=1 , n_3=# ; > plus | ; n_0=2 , n_1=# , n_2=# , n_3=1 ; >

The remaining 2 W bosons behave as (18.8)

| ; n_0=3 , n_1=0 , n_2=1 , n_3=1 ; >

The | ; n_0=3 , n_1=# , n_2=1 , n_3=1 ; > transforms into (18.9)

| ; n_0=2 , n_1=−1 , n_2=1 , n_3=1 ; >

The | ; n_0=2 , n_1=−1 , n_2=1 , n_3=1 ; > consists of a pair of correlated photons (18.10)

Suppose one observer measures 1 of the photons to be | ; n_0=1 , n_1=−1 , n_2=1 , n_3=0 ; > (or 2-polarization). Then, an observer measuring the other photon measures | ; n_0=1 , n_1=−1 , n_2=0 , n_3=1 ; > (or 3-polarization).

Fermion count is conserved. There is no need to describe a particle as moving backward in time.

The l-family field can participate in electron-positron pair production. For example, the reverse of above reaction can occur.

We discuss neutrino oscillations

People report that neutrinos oscillate. For example, a mu-neutrino can become a tau-neutrino. People report that the presence of mass triggers oscillations or enhances rates of oscillation. [Ref.18.1]

The next items describe a neutrino-oscillation event. (Possibly, the reaction requires additional reactants in order to conserve energy and momentum.)

- A mu-neutrino enters (18.11)
- A 3e4 state exists (here, $\upsilon>0$ provides an occupation number)
  $| ; n_0=1+\upsilon , n_1=-1 , n_2=\# , n_3=\# , n_4=\upsilon , n_5=\# ; >$
- A 4o(0) state exists (here, $n\geq0$ provides an occupation number)
  $| ; n_{-2}=\# , n_{-1}=\# , n_0=n , n_1=1+n ; >$
- A unit of 3e4 converts to a unit of 4o(0) (18.12)
  $| ; n_0=1+\upsilon , n_1=-1 , n_2=\# , n_3=\# , n_4=\upsilon , n_5=\# ; >$
  plus $| ; n_{-2}=\# , n_{-1}=\# , n_0=n , n_1=1+n ; >$
  $\rightarrow$
  $| ; n_0=\upsilon , n_1=-1 , n_2=\# , n_3=\# , n_4=\upsilon-1 , n_5=\# ; >$
  plus $| ; n_{-2}=\# , n_{-1}=\# , n_0=n+1 , n_1=2+n ; >$
- The neutrino increases its generation via an interaction with the unit of 4o(0) (18.13)
- The 4o(0) depopulates by 1 and becomes
  $| ; n_{-2}=\# , n_{-1}=\# , n_0=n , n_1=1+n ; >$
- A tau-neutrino exits (18.14)
- A 3e4 state exists (here, $\upsilon-1\geq0$ provides an occupation number)
  $| ; n_0=\upsilon , n_1=-1 , n_2=\# , n_3=\# , n_4=\upsilon-1 , n_5=\# ; >$
- A 4o(0) state exists (here, $n\geq0$ provides an occupation number)
  $| ; n_{-2}=\# , n_{-1}=\# , n_0=n , n_1=1+n ; >$

The value of υ correlates with the presence of mass. Based on concepts pertaining to lasing, the larger is υ, the higher is the probably (per unit time) of such an oscillation. The overall reaction does not change n for the 4o(0) state. People might say that the 4o(0) participates virtually.

Other similar reactions provide for oscillations between other pairs of types of neutrinos.

We illustrate mechanics of channels

The next items pertain to creation, from a ground state, of an even-polarized $e%& e-family boson. Here, the list % has $-n_1$ elements.

The number of channels available is $ (18.15)

- Aspect 1 (18.16)
  - For each of the $-n_1$ even-numbered $\chi\geq2$ for which $n_\chi=0$ (and $n_\chi\neq\#$), n_χ increases by 1
  - For 1 closed oscillator pair (with n_υ corresponding the pair's even-indexed oscillator),
    - The oscillator pair opens as $| n_\upsilon=0 , n_{\upsilon+1}=-1 >$
    - n_υ changes from 0 to n_1 (n_υ decreases)
- Aspect 2 (18.17)
  - n_0 increases by $-n_1$ (n_0 increases)
  - n_υ changes from n_1 to 0 (n_υ increases)
  - The n_υ-and-$n_{\upsilon+1}$ oscillator pair closes

The next items pertain.

In aspect 1, n_υ transits from non-negative to negative (18.18)
In aspect 2, n_υ transits from negative to non-negative (18.19)

Comments

We suggest research

SOR.18.1 Determine dependences of neutrino-oscillation rates on influences of mass or gravity.

We list references

Ref.18.1 J. Beringer et al. (Particle Data Group), *Phys. Rev. D86*, 010001 (2012). "13. Neutrino mass, mixing, and oscillations," page 46. (http://pdg.lbl.gov/2012/reviews/rpp2012-rev-neutrino-mixing.pdf)

Part 5 Baryonic matter, dark matter, and dark energy

Context

We note that people consider that most of the universe is not comprised of familiar matter

People use the term baryonic matter to discuss the type of matter people and equipment sense via photons. People say that energy associated with baryonic matter comprises less than about 5 percent of the energy in the universe.

People note that bases for the remaining energy have yet to be determined. People use the terms dark matter and dark energy to describe notions of energy correlating with other than baryonic matter.

We think IOM correlates with 2 bases for non-baryonic-matter stuff

Here, we discuss 2 possible bases for the remaining energy. IOM correlates with each basis. Each basis includes a concept for dark matter and a concept for dark energy. We think IOM correlates with the 2 bases not being mutually exclusive.

Core

We preview this part

One section discusses perspective about dark matter and dark energy.

One section discusses a possible explanation of dark matter and dark energy.

One section discusses another possible explanation of dark matter and dark energy.

One section discusses possibilities that the 2 proposed bases for dark matter and dark energy both exist.

Comments

We note that considering dark energy to be stuff may be non-traditional

People may consider an interpretation that dark energy is stuff to be non-traditional.

Section 19 Perspective about dark matter and dark energy

Abs.19.1 We note and interpret observations regarding effects of dark matter and dark energy.

Context

We note that people interpret observational results as correlating with dark matter and dark energy

People interpret observations of extra-terrestrially generated photons as correlating with the existence of dark matter and dark energy.

We summarize some results and thinking regarding dark matter and dark energy

Here, we summarize some results regarding dark matter and dark energy. And, we note some thinking regarding observations people correlate with the existence of dark matter and dark energy.

Core

We note some astrophysical observations

People discuss non-uniformities in CMB (cosmic microwave background) radiation. People interpret some of the non-uniformities as correlating with the existence of things people call dark matter and dark energy. People have yet to settle on the nature of dark matter or the nature of dark energy.

Traditional conjectures vary regarding the nature of dark matter. Traditional physics seems to include the following statements. The systems people use to do experiments and make observations consist of baryonic matter. Dark matter and baryonic matter interact via gravity. Baryonic-matter systems do not detect light emitted recently by dark matter. Perhaps, experiments can detect dark matter.

People infer the existence of dark matter. Relevant inferences come from observations of baryonic matter and photons. People discuss possibilities that dark matter produces its own type of photons. People opine that perhaps dark matter is too cold to produce observable photons. People posit the existence of dark matter to explain aspects of the motions and shapes of some baryonic-matter galaxies. People posit that dark matter affects such motions and shapes via gravity. [Ref.19.1]

People estimate ratios of contributions to the density of the universe of baryonic matter, dark matter, and dark energy. People base estimates on data about CMB. Traditional physics considers that observed CMB photons have existed since near the time of the big bang. Physics considers that before $10^{5.6}$ years after the big bang, such photons interacted significantly with ionized plasma. Around $10^{5.6}$ years after the big bang, the plasma ceased to be ionized. Much CMB radiation travels today.

Data suggest late-time (or secondary) anisotropies. Late-time refers to any time after the above-mentioned plasma ceased to be ionized. Relevant processes for generating non-uniformities in CMB may feature photon scattering by free electrons (Thompson scattering), scattering by clouds of high-energy electrons (Compton scattering by hot electron gases), and frequency-shifting because of changing gravitational fields (integrated Sachs-Wolfe effect). [Ref.19.2]

Observations pertaining to just after $10^{5.6}$ years after the big bang fail to detect a significant density of the universe for dark energy.

The next items note interpretations of observations pertaining to recent times. [Ref.19.3]

A ratio of densities for baryonic matter and dark matter is $\sim 1 : 5.5$	(19.1)
A ratio of densities for baryonic matter and dark energy is $\sim 1 : 13.9$	(19.2)

Comments

We note some possible interpretations of observations

Perhaps it is fair to say that observational and experimental results do not contradict the next items. We consider the possibility that dark energy is stuff (like baryonic matter and dark matter).

No interactions take place between dark-matter photons (if such exist) and baryonic matter	(19.3)
No interactions take place between baryonic-matter photons and dark matter	(19.4)

Gravity relevant to baryonic matter interacts with at least approximately 6 times as much density-of-the-universe stuff as baryonic-matter photons interact with (19.5)
Gravity relevant to baryonic matter does not interact with dark-energy stuff (19.6)
Gravity (if such exists) relevant to dark-energy stuff does not interact with baryonic matter (19.7)

We suggest research

SOR.19.1 Find or rule out (to some confidence level) evidence (other than currently assumed evidence) of effects on baryonic matter of dark matter.
STR.19.1 How best should people try to directly detect matter not associated with baryonic matter?

We list references

Ref.19.1 J. Beringer et al. (Particle Data Group), *Phys. Rev. D86*, 010001 (2012). (http://pdg.lbl.gov/2012/reviews/rpp2012-rev-dark-matter.pdf)
Ref.19.2 J. Beringer et al. (Particle Data Group), *Phys. Rev. D86*, 010001 (2012). (http://pdg.lbl.gov/2012/reviews/rpp2012-rev-cosmic-microwave-background.pdf)
Ref.19.3 Mark Peplow, Planck telescope peers into primordial universe, *Nature News*, Nature Publishing Group, March 21, 2013. (http://www.nature.com/news/planck-telescope-peers-into-primordial-universe-1.12658)

Section 20 Dark matter and dark energy, possibility 1 - ensembles

Abs.20.1 A possible symmetry points to 48 ensembles (instances of Standard Model particles).
Abs.20.2 Adequately physics-savvy beings in any 1 of the ensembles could deduce the existence of 5 dark matter ensembles and 18 dark energy ensembles.

Context

We think it possible that dark matter and dark energy are similar to baryonic matter

Perhaps, people can emphasize the possibility that dark matter and dark energy have similarities to baryonic matter.

We discuss a possible basis for dark matter and dark energy

We interpret possible IOM symmetries as suggesting possible bases for dark matter and dark energy.

Core

We identify a possible symmetry pertaining to the s- and e-families

The next items show numbers of matches between groups of s- and e-family oscillators and symmetry labels. We use ó to denote the number of oscillators for a row in the chart. For that row, we assign ó labels. ó! provides the number of ways to match each label noted in a row with each oscillator noted for the row. We assign 1 factor of 1/2 to compensate for a redundancy in choices. For the o- and w-families, exchanging labels for n_{-2} and n_{-1} correlates with exchanging labels for n_2 and n_3. For the o- and w-families, such an exchange reverses the signs of charges. We extend that correlation to the s- and e-families. Items including and following item (20.8) further explain the numbers of labels.

χ for oscillators n_χ	ύ for symmetry labels $sl_{ύ}$	Matches	Factor	Product (top down)	Product (bottom up)	(20.1)
−2. −1, 0	−2. −1, 0	6			48	(20.2)
1	1	1				(20.3)
2, 3	2, 3	2	1/2	6		(20.4)
4, 5	4, 5	2		12	8	(20.5)
6, 7	6, 7	2		24	4	(20.6)
8, 9	8, 9	2		48	2	(20.7)

The next items provide further explanation.

People may correlate matches regarding sl_{-2} , sl_{-1} , and sl_0 with assignments of 3 color charges (20.8)
People may correlate other matches with notions of parity (20.9)
For example, for n_4 and n_5, a reversal of sl_4 and sl_5 correlates with a different parity with respect to a non-changing assignment of sl_{-4} and sl_{-3} to n_{-4} and n_{-3} (20.10)

We apply the symmetry and model 48 instances of Standard Model particles

The next items interpret the matches. For example, for each instance of gravitons, 6 instances of Standard Model particles exist. In total, 48 instances of Standard Model particles exist.

Gss.20.1 People can use the matching of oscillators and symmetry labels to characterize a symmetry. (20.11)
For each instance of 1e%& particles, 2 instances of 2e%& particles exist (20.12)
For each instance of 2e%& particles, 2 instances of 3e%& particles exist (20.13)
For each instance of 3e%& particles, 6 instances of Standard Model particles (including 4e2&) exist (20.14)

The next item reviews math leading to the number 48.

$$48 = 6 \times 2 \times 2 \times 2$$
$$6 = 6 \times 1 \times 2 \times (1/2)$$
(20.15)

Comments

We define the term ensemble and we define related terms

The next item defines the term ensemble.

The term ensemble denotes an instance of particles similar to Standard Model particles. (20.16)

The next items group and label ensembles. We state these results relative to the baryonic-matter ensemble. A column notes numbers of ensembles within a group. The next column shows how we compute the numbers. The next column shows how we calculate inputs to the computations. In a row in that column, a first calculation ties to item (20.15). In that row, the second calculation ties to numbers of ensembles we compute for the rows above that row.

Traditional name	Our name	Number of ensembles	Some math	Some more math	(20.17)
Baryonic matter	Ensemble@1	1	1	1	(20.18)
Dark matter	Ensembles@5	5	5=6−1	6=1×6 1=1	(20.19)
Dark energy	Ensembles@6#2	6	6=12−6	12=1×6×2 6=1+5	(20.20)
Dark energy	Ensembles@12#2	12	12=24−12	24=1×6×2×2 12=1+5+6	(20.21)
-	Ensembles@24#2	24	24=48−24	48=1×6×2×2×2 24=1+5+6+12	(20.22)

The next items provide notation. Here, ∪ denotes the union of 2 sets.

ensembles@6#1 = ensemble@1 ∪ ensembles@5 (20.23)
ensembles@12#1 = ensembles@6#1 ∪ ensembles@6#2 (20.24)
ensembles@18 = ensembles@6#2 ∪ ensembles@12#2 (20.25)
ensembles@24#1 = ensembles@12#1 ∪ ensembles@12#2 (20.26)
= ensembles@6#1 ∪ ensembles@18
ensembles@47 = ensembles@5 ∪ ensembles@18 ∪ ensembles@24#2 (20.27)
ensembles@48 = ensembles@24#1 ∪ ensembles@24#2 (20.28)

We interpret results

The next item applies.

Gss.20.2 People can consider any ensemble to be (relative to itself) its own ensemble@1. (20.29)

The next items apply.

Possibly, any ensemble contains physics-savvy beings (20.30)
Adequately physics-savvy beings would consider that, relative to the beings' ensemble, … (20.31)
- 5 ensembles of what the beings consider dark matter exist
- 18 ensembles of what the beings consider dark energy exist

Dark matter correlates with a reciprocal relationship between 2 ensembles (20.32)
- Such ensembles share an instance of 3e4& forces

Dark energy correlates with a reciprocal relationship between 2 ensembles@24#1 ensembles (20.33)
- Such ensembles do not share an instance of 3e%& forces

We know of no reasons to rule out the next items.

Effects of ensembles@5 have been observed (20.34)
- People attribute about a quarter of the density of the universe to dark matter
- Unusual shapes of some ensemble@1 galaxies may be caused by interactions with ensembles@5 galaxies
- Interpretations of CMB observations are not inconsistent with the existence of ensembles@5

Effects of ensembles@18 have been observed (20.35)
- People attribute much of the density of the universe to dark energy
- People state that dark energy does not interact gravitationally with ensemble@1
- Effects of ensembles@18 on CMB have yet to reach their ultimate amounts

We note reasons why the observed ratios need not be 1:5:18

The next items provide possible reasons why observed ratios of densities of ensemble@1, ensembles@5, and ensembles@18 need not be 1:5:18.

Ensembles may not have been created with equal densities (20.36)
Ensembles may have evolved at different rates (20.37)
The universe has not existed long enough (20.38)
- For example, there has not yet been enough time for 1e2468& to transfer to ensemble@1 enough effects of (or information about) non-uniformities in ensembles@18

Observations pertain to too small a portion of the universe (20.39)
Forms of dark matter other than ensembles@5 also pertain (20.40)
Forms of dark energy other than ensembles@18 also pertain (20.41)
Ensembles@5 do not exist, but other stuff provides for effects we attribute to ensembles@5 (20.42)
Ensembles@18 do not exist, but other stuff provides for effects we attribute to ensembles@18 (20.43)

We discuss time delays in the effects of ensembles@18 on CMB

We explain the seeming delay regarding observing the 18 in 1:18 regarding the ratio of the observed density for ensemble@1 to the observed density for ensembles@18. [(19.2) and (20.38)]

Analyses of CMB data determine, in effect, aspects of clumping. We use the term clumping to denote non-homogeneity in the distribution of energy or matter. Clumps include atoms, planets, stars, solar systems, galaxies, galactic clusters, clouds of electrons, and so forth.

For people to observe a ratio of 1:18, people would need to look at data pertaining to a time after the big bang for which all of the next items have substantially taken place.

Ensembles@12#2 has clumped (20.44)
1e%& interactions have induced clumping in ensembles@12#1 based on clumping in ensembles@12#2 (20.45)

1e%& interactions and 2e%& interactions have induced clumping in ensembles@6#1 based on clumping in ensembles@6#2 and ensembles@12#2 (20.46)

- 2e%& interactions link ensembles@6#1 and ensembles@6#2
- 2e%& interactions do not link ensembles@6#1 and ensembles@12#2

1e%& interactions, 2e%& interactions, and 3e%& interactions have induced clumping in ensemble@1 based on clumping in ensembles@5, ensembles@6#2, and ensembles@12#2 (20.47)

- 3e%& interactions link ensemble@1 and ensembles@5
- 3e%& interactions do not link ensemble@1 and ensembles@18

CMB photons (some of the 4e2& associated with ensemble@1) have reacted to the following (20.48)

- Clumping of objects in ensemble@1
- Effects of clumping on the gravitational field (and other fields) germane to CMB photons

Thus, it takes time for the observed (based on CMB) density ratio of ensemble@1 to ensembles@18 to increase from 1:(~0) to 1:18.

We discuss a possible way to observe that 48, not just 24, ensembles exist

The next item suggests that people might be able to observe effects of ensembles@24#2. Such a possible asymmetry likely, but not necessarily, would correspond to asymmetry in the spatial distribution of ensemble@1. Such asymmetry likely would have been formed early in the history of the universe.

Gss.20.3 Asymmetry in the spatial distribution of ensembles@24#1 might correlate with the existence of ensembles@24#2. (20.49)

We suggest research

SOR.20.1 Determine the extents to which fermion properties vary by ensemble.

SOR.20.2 Determine limits on similarities and differences between ensembles@5 matter and ensemble@1 matter.

SOR.20.3 Use data about ensemble@1 to infer w-, h-, and o-family interaction strengths and/or big-bang phenomena related to w-, h-, and o-family-mediated interactions.

SOR.20.4 Determine the extents to which fields related to various 3e%&, 2e%&, and 1e%& bosons have impacted CMB radiation.

SOR.20.5 Determine the extent to which observations (for example, about the uniformity of distribution of ensembles@24#1 stuff) cannot be explained by traditional physics and/or work in this paper directly related to ensembles@24#1. Estimate the extent to which people might attribute any such unexplained phenomena to the presence of ensembles@24#2 (for example, via effects early in the big bang).

SOR.20.6 Detect or rule out (to some confidence level) effects on ensembles@24#1 of ensembles@24#2.

SOR.20.7 Determine the extent to which ensembles contribute equally to actual densities of the universe. (Here we differentiate actual from inferred {for example, inferred via CMB data}.)

STR.20.1 How might people detect or rule out (to some confidence level) the existence of ensembles@24#2?

STR.20.2 To what extent should theories related to gravitation, curvature of space time, general relativity, and so forth be reexamined because of the existence of 4 similar or 8 total forms of gravity (3e4&)?

STR.20.3 To what extent does trying to account, in the Einstein field equations, for dark energy via the cosmological constant suffice?

STR.20.4 To the extent unchanged Einstein field equations pertain, to what extent does the cosmological-constant term vary (as a function of position in space time) because of clumping of dark-energy stuff?

STR.20.5 For each of the various eras ($Z_{\circlearrowleft}$, per Section 12), which interactions most affect neutrinos?

STR.20.6 To what extent might people use the term universe to denote ensembles@48 or just ensembles@24#1?

Section 21 Dark matter and dark energy, possibility 2 - basic fermions having S≥3/2

Abs.21.1 Possible S=3/2 and S=7/2 q-family basic fermions may provide bases for dark matter and dark energy.

Context

We note that we identify above possible yet-to-be-detected fermions

Our work regarding the q-family suggests possible as-yet-not-detected particles.

We discuss a possible basis for dark matter and dark energy

We interpret the possible existence of S=3/2 and S=7/2 q-family basic fermions as suggesting possible bases for dark matter and dark energy.

Core

We note that S≥3/2 basic fermions could contribute to dark matter and dark energy

The next item correlates with S=3/2 q-family basic fermions being candidates for dark matter. Here, we use the term directly to parallel considerations regarding neutrinos. Neutrinos have no charge. But, people say that neutrinos can interact indirectly with photons. [Compare STR.12.2.]

> Gss.21.1 Q-family basic fermions for which $n_2=n_3=-1$ (with respect to interactions with the w-, h-, and o-families) do not interact directly with 4e2& (photons). (21.1)

The next item correlates with S=7/2 q-family basic fermions being candidates for dark energy.

> Gss.21.2 Q-family basic fermions for which $n_2=n_3=n_4=n_5=n_6=n_7=-1$ (with respect to interactions with the w-, h-, and o-families) do not interact directly with any e-family members other than 1e%& for which 8є%. (21.2)

Comments

We discuss the possibility that S≥3/2 fermions provide for all dark matter and dark energy

We know of no evidence suggesting S≥3/2 q-family fermions could not provide bases for all dark matter and dark energy. If S≥3/2 q-family fermions provide bases for all dark matter and dark energy, of the possible 48 ensembles [Section 20], only ensemble@1 exists.

For the possibility that S=3/2 fermions provide all dark matter and S=7/2 fermions provide all dark energy, we offer no models for ratios of densities of baryonic matter, dark matter, and dark energy.

We note possible transitions between forms of stuff

We interpret items (11.52) and (11.75) as indicating possibilities for S≥3/2 fermions to transform into S=1/2 fermions.

We suggest research

STR.21.1 Estimate ratios of densities of the universe for baryonic matter, dark matter, and dark energy, assuming that S≥3/2 basic fermions provide all dark matter and dark energy.

Section 22 Dark matter and dark energy - hybrid model

Abs.22.1 The 2 bases (ensembles and S≥3/2 basic fermions) for dark matter and dark energy can both apply.

Context

We note that both types of dark matter and both types of dark energy might exist

We know of no evidence that nature cannot have both S≥3/2 basic fermions and more than 1 ensemble.

We discuss the possibility that both types of non-baryonic stuff exist

We discuss the possibility that nature includes both S≥3/2 basic fermions and more than 1 ensemble.

Core

We discuss possible associations of S=3/2 and S=7/2 basic fermions with ensembles

If nature includes both forms of dark matter (ensembles@5 and S=3/2 q-family basic fermions) and both forms of dark energy (ensembles@18 and S=7/2 q-family basic bosons), the question arises as to the extent S≥3/2 basic fermions correlate with specific ensembles.

The next items present 2 possibilities. We base the first item on IOM(−8;−8,9;9) representations for which n_4 or $n_5 = -2$ (for S=3/2) and n_8 or $n_9 = -2$ (for S=7/2). We base the second item on the notion that a smaller value of Ω correlates with a smaller or not greater range of ensembles. For S=3/2 and S=7/2, smaller Ω correlates with larger |Ω|.

Ensembles share S≥3/2 basic fermions (22.1)
- For ensembles@6#1, 1 set of S=3/2 basic fermions exists
- For each peer of ensembles@6#1, 1 set of S=3/2 basic fermions exists
- For ensembles@24#1, 1 set of S=7/2 basic fermions exist
- For ensembles@24#2, 1 set of S=7/2 basic fermions exist

Each ensemble has its own S≥3/2 basic fermions (22.2)

Comments

We discuss possible conclusions

One of the next items pertains, assuming that no third basis for dark matter and dark energy exists. We know of no observations that rule out any such alternative.

No S≥3/2 basic fermions exist (22.3)
- A basis for approximately 1:5:18 ultimate ratios of baryonic matter, dark matter, and dark energy pertains
  - Exact 1:5:18 ratios need not be achievable or observed, per items starting with item (20.36)

No ensemble in ensembles@47 exists (22.4)
- S≥3/2 basic fermions provide the bases for dark matter and dark energy
- We do not offer a basis for predicting ultimate (or current) density of the universe ratios for baryonic matter, dark matter, and dark energy

Ensembles@24#1 exists, S≥3/2 basic fermions exist, and ensembles share S≥3/2 basic fermions (22.5)
- A basis for 1:5+:18+ ultimate ratios of baryonic matter, dark matter, and dark energy pertains
  - Perhaps, the amount that 5+ exceeds 5 correlates with S=3/2 basic fermions shared by the 6 ensembles in ensembles@6#1

Ensembles@24#1 exists, S≥3/2 basic fermions exist, and ensembles do not share S≥3/2 fermions (22.6)
- S≥3/2 basic fermions add to properties (such as property-3) for their respective ensembles
- A basis for approximately 1:5:18 ultimate ratios of baryonic matter, dark matter, and dark energy pertains
  - Exact 1:5:18 ratios need not be achievable or observed, per items starting with item (20.36)

We suggest research

SOR.22.1 Perform observations or experiments sufficient to allow people to decide among at least 4 hypotheses regarding S≥3/2 fermions, ensembles, and non-baryonic matter. [items (22.3), (22.4), (22.5), and (22.6)]

STR.22.1 Review observational data and calculations of the density of baryonic matter (equivalent to a mass density of approximately 9.9×10^{-30} g/cm^3, which is similar to mass density of 5.9 protons per cubic meter) to find a basis for determining the extent to which S≥3/2 fermions might pertain. [Ref.22.1]

We list references

Ref.22.1 Wilkinson Microwave Anisotropy Probe,
http://wmap.gsfc.nasa.gov/universe/WMAP_Universe.pdf

Part 6 Perspective

Context

We note that above we leave some topics not addressed and others not summarized

Work above does not address some possibly relevant topics. Work above does not contain a summary.

We anticipate providing perspective and providing summary material

We can add perspective and provide some summary material.

Core

We preview sections in this part

One section discusses possible relationships between work in this paper and various physics topics. One section summarizes progress this paper may represent.

Section 23 Relationships between this work, other models, and physics

Abs.23.1 We correlate IOM with traditional physics topics, models, and theories.

Context

We note that above we do not address possibly correlated aspects of physics

We tend to avoid, above, extending discussion so as to include nuclear physics and other possibly relevant physics topics.

We note that work above may present opportunities in fields not addressed above

People may find implications and uses - in various areas of physics and applied physics - of work in or stemming from this paper. We can provide some possible examples.

Core

We posit relationships between IOM and traditional physics topics

The next items come from work in this paper.

IOM(−2;−2,3;9) correlates with all known basic particles (23.1)
IOM(−8;−8,9;9) indicates possibilities for basic particles people have not discovered (23.2)

The next items posit relationships between traditional physics (models and theories) and IOM. For the relativity items, traditional notes and our notes address quantum mechanics people relate to non-quantum theories.

Theory	Traditional notes	Our notes	IOM numbers	Notes about IOM version	(23.3)
Special relativity	Does not predict a spectrum of basic non-zero mass particles	Does not predict masses for basic particles other than photons	0;0,3;3	-	(23.4)
General relativity	Possibly, no established quantum version exists	Might not predict masses for basic particles other than photons and gravitons	0;0,5;5	Includes a basis for a quantum version	(23.5)
Standard Model	Does not include gravity	Does not include basic particles for which S>1	−2;−2,3;3	Does not include basic particles for which S>1	(23.6)
Our model	-	-	−8;−8,9;9	May correlate with the observable universe (... and ...) Includes basic particles for which 0≤S≤4	(23.7)

Comments

We discuss compatibility between relativity and IOM

The next items pertain.

Special relativity and general relativity correlate with $\Omega \geq 0$ (23.8)

For $b_2 \leq -2$, IOM(b_1;b_2,b_3;b_4) includes $\Omega < 0$ (23.9)

Gss.23.1 People might consider that incompatibilities exist between $\Omega < 0$ and each of traditional special relativity and traditional general relativity. (23.10)

We discuss some aspects of traditional physics

Possibly, people will use IOM to gain new insight regarding the next items.

Topic in traditional physics	Notes	(23.11)
• Nuclear shell model	• People might use IOM to hone or supplant the nuclear shell model	(23.12)

Topic in traditional physics	Notes	(23.11)
• Dark energy	• May be comprised of stuff • May not be the cause of current increases in the rate of expansion of the universe • May be involved in interactions that lead to phenomena people correlate with the observed increase in rate	(23.13)
• Higgs field	• To the extent people say such has a role in providing for non-zero masses, possibly people should say that such has a role in providing for non-zero charges and non-zero magnetic moments	(23.14)
• Planck length	• Correlates with confluence of 2 series (the photon-graviton series and the maximal-% series) of e-family members • Correlates with uncertainties related to attempts to make precise measurements • May not correlate with any granularity of space time	(23.15)
• Speed of light	• May not be a limit regarding $\Omega<0$ phenomena	(23.16)
• Zero-point energy	• People may find using Œ=0 obviates concerns about (otherwise) possibly infinite zero-point energy associated with a potentially infinite number of photon ground states	(23.17)
• Number of independent physical constants	• People may say that use of IOM helps reduce this number	(23.18)
• Schwarzschild (black hole) radius	• People may find use (other than regarding black holes) for this concept	(23.19)
• Uncertainty	• People may find value in considering that invariant states of non-zero-mass basic particles have zero size • People may continue to associate non-zero uncertainty with superposition of states people associate with kinematics	(23.20)
• Origin of the universe; multiple universes	• Perhaps people will find benefit in exploring IOM(b_1;b_2,b_3;b_4) for which $b_2<-8$ or $b_3>9$	(23.21)

Topic in traditional physics	Notes	
• (Possible) Smallest non-zero quanta of charge, property-3, or magnetic moment	• Perhaps people will gain insight by exploring $\Omega<0$ physics	(23.11) (23.22)

We note that some physics constants approximate some well-know numbers

Perhaps people will find the existence of following approximations useful when people try to develop new models or more-fundamental theory.

$$\beta \sim (e^e)^3 \quad (23.23)$$
$$\alpha \sim \exp(-(2/3)e^2) \quad (23.24)$$

We suggest research

SOR.23.1 Detect or rule out (to some confidence level), say for 3e4& and 2e6&, fields analogous to the magnetic field people associate with 4e2&.
SOR.23.2 To what extent might non-4e2% e-family fields analogous to the magnetic field correlate with jets associated with quasars or with other phenomena?
STR.23.1 To what extent would people benefit by correlating IOM(−1;−1,2;2) with Dirac-equation spinors? (Each has 4 components. Here, $D_e=D_p=2$.)
STR.23.2 To what extent would people benefit by considering Q'=0 to be a charge, such as Q'=−1, and not just a lack of charge?
STR.23.3 To what extents might people consider that traditional-physics special relativity and traditional-physics general relativity do no precisely apply at distances for which o- and s-family interactions have significant influence?
STR.23.4 To what extents might people consider that traditional-physics special relativity and traditional-physics general relativity do no precisely apply (for a non-zero-mass system) at distances smaller than roughly a system's spin/mass length [(6.15)]?
STR.23.5 Estimate probabilities of significant perturbations to the evolution of earth, to planetary motion, or to other solar-system phenomena, because of dark-matter stuff.
STR.23.6 To what extent might people, based on this work, change estimates of the number of independent physical constants?

Section 24 Summary and concluding remarks

Abs.24.1 We list advances this paper may provide.
Abs.24.2 We provide possibly useful perspective from other work and about future opportunities.

Context

We note that new approaches can yield breakthroughs

We have heard that Albert Einstein said something like that people cannot solve problems with the thinking that led to the problems.

The next items sketch breakthroughs regarding modeling perceived motions of extra-terrestrial objects.

Actions	Notes	(24.1)
• People developed models based on epicycles	• People benefitted from the models • The models grew more complex	(24.2)
• Copernicus encouraged people to think of the sun as a central body	• People benefitted from better models • People considered fewer epicycles	(24.3)
• Kepler suggested elliptical orbits	• People gained a more useful basis for calculations • People deemphasized epicycles	(24.4)
• Newton suggested gravity as a basis for elliptical orbits	• People's abilities to model and predict motions grew	(24.5)
• Einstein embedded a theory of (non-quantum) gravity within general relativity	• People's abilities to model astrophysical phenomena grew	(24.6)

We note possible parallels between scenarios

The next items posit statements people might make.

The Standard Model and the epicycle approach share attributes of being patchworks (24.7)
IOM features guesses and as-yet-not-integrated aspects (24.8)
People might say that neither the Standard Model nor IOM qualifies as a theory (24.9)
People may be able to develop theory (for topics people use the Standard Model and/or IOM to attempt to address) that includes or produces adequately useful models (24.10)

Core

We estimate the extent to which this research meets some needs

Section 1 and Section 2 list needs people seek to meet via mathematical models.

The next items provide estimates as to the extent to which work in this paper addresses needs Section 1 lists.

Needs not met via the Standard Model	Possible advances	(24.11)
Provide models people can use to ...	Work in this paper may ...	(24.12)
• List possible basic particles that have not been observed	• Fulfill this need	(24.13)
• Describe quantum gravity	• Fulfill this need	(24.14)

Needs not met via the Standard Model	Possible advances	(24.11)
Provide models people can use to ...	Work in this paper may ...	(24.12)
• Unify quantum gravity and quantum electromagnetism	• Fulfill this need	(24.15)
• Explain dark matter	• Describe 2 possibilities (1 in detail and 1 in not so much detail) for not-mutually-exclusive types of dark matter	(24.16)
• Explain dark energy	• Describe 2 possibilities (1 detail and 1 in not so much detail) for not-mutually-exclusive types of dark energy	(24.17)
• Explain changes in the rate of expansion of the universe	• Point to forces that regulate the rate	(24.18)
• Explain baryon asymmetry (matter/antimatter imbalance)	• Point to interactions that led to this phenomenon	(24.19)
• Explain the sizes of some symmetry violations (P, CP, ...)	• Point to interactions that include and augment interactions people correlate with Standard Model physics	(24.20)
• Explain neutrino oscillations	• Illustrate an interaction that facilitates this phenomenon	(24.21)
• Predict neutrino masses	• Point to a few candidate masses	(24.22)
• Explain the number, 3, of generations of fermions	• Fulfill this need	(24.23)
• Determine whether magnetic monopoles ever existed	• Rule out fermion magnetic monopoles	(24.24)
• Address the zero-point energy of the vacuum	• Provide a law (Œ=0) that obviates this concern	(24.25)
• Interrelate masses of basic particles other than charged leptons	• Point to approximate relationships between masses of basic bosons • Point to approximate relationships between masses of S=1/2 basic fermions	(24.26)

The next items provide estimates as to the extent to which work in this paper addresses needs Section 2 lists.

Needs met via the Standard Model		
Provide models people can use to ...	Work in this paper may ...	(24.28)
• List basic particles that have been observed	• Fulfill this need	(24.29)
• Describe and unify the electromagnetic, weak, and strong interactions	• Fulfill this need	(24.30)

(24.27)

Comments

We think the IOM approach has promise

We hope and think these things, regarding at least items following item (24.11) and items following (24.27). Here, the approach denotes generally work in this paper and specifically IOM.

- People will find value in the approach and results from the approach (24.31)
- People will fix any flaws in the approach and related results (24.32)
- People will find better, more-unified bases for aspects of the approach (24.33)
  - People will develop theory that people can use to produce models
  - Such theory will obviate seeming needs for various guesses
- People will conduct observations and experiments based on people's knowledge of the approach and on results from the approach (24.34)
- People will determine or verify physical numbers, based on such results (24.35)
- People will find better ways to present such theory, the approach, and related results (24.36)

We suggest research

STR.24.1 Develop more-coherent or more-compact (than this paper shows) models that produce or improve on results this paper offers.

STR.24.2 Develop theory people can use to produce models (such as this paper shows or STR.24.1 suggests).

STR.24.3 To what extent can people benefit by developing hybrid models that include aspects of IOM and the Standard Model?

STR.24.4 To what extent can people develop theory that generates adequately useful, possibly hybrid models (that include aspects of IOM and the Standard Model)?

STR.24.5 For what other applications might people use IOM?

Part 7 Appendices

Context

We note that above we do not compile lists of some types of material

We do not include above compilations people may find useful.

We note services this part provides

Here, we compile lists (such as of guesses and of references) tabulated from various sections.

Core

We preview sections in this part

One section brings together various lists of elements in this paper. The elements are section abstracts, guesses, suggested observational and experimental research, and suggested theoretical research.

One section brings together a list of references this paper cites.

Section 25 Compendia of section abstracts, guesses, and suggested research

We list statements in section abstracts

Abs.1.1	Traditional mathematical models do not adequately correlate with physics observations.
Abs.2.1	Quantum isotropic harmonic oscillator methods (IOM) may adequately correlate with physics observations for which traditional models do not correlate.
Abs.3.1	Sections in this paper compile lists of inputs to and results from research this paper describes.
Abs.4.1	We introduce IOM (quantum isotropic harmonic oscillator methods, math, and models).
Abs.5.1	We focus on IOM for which $\Omega = \pm S(S+1)$, with $S=\text{spin}/\hbar$, 2S being an integer, and $0 \le S \le 4$.
Abs.6.1	The mass of a tauon may equal a number computed from 4 physics constants.
Abs.6.2	We define a series of lengths, including the Planck length, based on 4 physics constants.
Abs.6.3	We note invariant properties of basic particles and of objects.
Abs.7.1	A catalog of families of basic particles correlates with possible yet-to-be-discovered particles.
Abs.8.1	The e-family includes photons, gravitons, and 2 other zero-mass basic bosons.
Abs.8.2	The e-family includes coherences of the family's 4 basic bosons.
Abs.8.3	Each e-family member has 2 modes (polarizations).
Abs.8.4	Each of the 4 e-family basic bosons mediates a force with spatial dependence R^{-2}.
Abs.8.5	E-family coherences provide forces with spatial dependences of R^{-4}, R^{-6}, and R^{-8}.
Abs.8.6	This paper may provide a way to avoid dealing with infinite photon ground-state energy.
Abs.9.1	Non-zero-mass basic bosons include the w-family (Z, W^-, and W^+), the Higgs boson, and o-family bosons (for which $\Omega<0$).
Abs.10.1	S-family bosons provide for gluons for each of 2 sets of 3 color charges.
Abs.11.1	IOM correlates with leptons, quarks, and related fields.
Abs.11.2	IOM correlates with possible basic fermions with S=3/2 and with S=7/2.

Abs.11.3 One IOM interpretation correlates with each n-type (or neutrino-like) basic fermion being its own antiparticle. One IOM interpretation correlates with each basic fermion being distinct from its antiparticle.
Abs.11.4 Each q- or l-family particle is a member of a 3-generation trio.
Abs.12.1 E-family coherences provide for changes in the rate of expansion of the universe.
Abs.13.1 Lasing of o-family particles provided key effects leading to matter/antimatter imbalance.
Abs.13.2 O-family particles provide for CPT-related symmetry violations.
Abs.13.3 O-family particles with $S \geq 2$ close gaps between magnitudes of violations people estimate via the Standard Model and magnitudes people measure.
Abs.13.4 E-family coherences provide for phenomena people attribute to axions.
Abs.14.1 IOM correlates with kinematics of e-family and s-family bosons.
Abs.15.1 IOM correlate with relative masses for w- and h-family bosons.
Abs.15.2 IOM may correlate with masses for o-family bosons.
Abs.16.1 A formula approximates masses of quarks and charged leptons.
Abs.16.2 The formula for masses of charged basic fermions may provide masses for neutrinos.
Abs.16.3 IOM may approximately correlate with relative masses for quarks and charged leptons.
Abs.17.1 Formulas provide approximate relative strengths for interactions mediated by e-family basic bosons and some e-family coherences.
Abs.18.1 We illustrate interactions involved in fermion-anti-fermion annihilation.
Abs.18.2 We illustrate interactions involved in neutrino oscillations.
Abs.18.3 We illustrate mechanics of channels.
Abs.19.1 We note and interpret observations regarding effects of dark matter and dark energy.
Abs.20.1 A possible symmetry points to 48 ensembles (instances of Standard Model particles).
Abs.20.2 Adequately physics-savvy beings in any 1 of the ensembles could deduce the existence of 5 dark matter ensembles and 18 dark energy ensembles.
Abs.21.1 Possible $S=3/2$ and $S=7/2$ q-family basic fermions may provide bases for dark matter and dark energy.
Abs.22.1 The 2 bases (ensembles and $S \geq 3/2$ basic fermions) for dark matter and dark energy can both apply.
Abs.23.1 We correlate IOM with traditional physics topics, models, and theories.
Abs.24.1 We list advances this paper may provide.
Abs.24.2 We provide possibly useful perspective from other work and about future opportunities.

We list guesses

Gss.4.1 For an edge case with -2ν an even positive integer, 1 type-1 solution exists.
Gss.4.2 For an edge case with -2ν an odd positive integer, 3 orthogonal type-1 solutions exist.
Gss.5.1 For basic elementary particles, subsets of IOM(−8;−8,9;9) pertain. Parameters b_2 and b_3 in IOM(b_1;b_2,b_3;b_4) correlate with spin. $\nu=-1$ correlates with basic bosons. $\nu=-3/2$ correlates with basic fermion particles. $\nu=-1/2$ correlates with fermion fields.
Gss.6.1 $\beta' = \beta$.
Gss.6.2 We attach significance to $\lambda_\$$ for which a particle property has an exponent $\gamma=0$.
Gss.6.3 Regarding λ_5, people can consider q_e to be a particle property for which $|q_e|^0$ pertains.
Gss.6.4 The particle properties spin/$\hbar$ ($S=1/2$), charge (q_e), mass (m_e). magnetic moment ((g_S)$\hbar$/2, with $g_S=2$), fermion count (1), and handedness/chirality (left) characterize an electron.
Gss.6.5 The properties spin/$\hbar$, charge, property-3, magnetic moment, fermion count, and net handedness characterize an object.
Gss.8.1 For a basic boson, $n_1<0$ correlates with the boson's having no mass and $n_1 \geq 0$ correlates with the boson's having non-zero mass.

Gss.8.2 For a basic particle with even 2S, with $D_p>1$, and with $n_1<0$, the force imparted between 2 non-overlapping objects scales as $R^{ύ}$. Here, $ύ=2n_1$. Here, R denotes the distance between the center of property of one object and the center of property of the other object.
Gss.8.3 $D_e=1$, $D_p=5$ solutions provide a model for gravitons.
Gss.8.4 In the expression $(4/3)(\beta^6)^2 = \{(q_e)^2/(4\pi\varepsilon_0)\} / \{G_N(m_e)^2\}$, the leftmost exponent 2 represents the number of vertices in a Feynman diagram, β^6 represents the ratio of strengths per channel for electromagnetism and gravity (for an interaction between 2 electrons), 4 represents the number of channels for a photon, and 3 represents the number of channels for a graviton.
Gss.8.5 A channel corresponds to a closed harmonic-oscillator pair.
Gss.8.6 $e%& forces for which % contains a 6 couple to magnetic moment.
Gss.8.7 $e%& forces for which % contains an 8 couple to property-1.
Gss.8.8 For e-family basic bosons, 2S = the maximum χ for which the χ-and-$(\chi+1)$ pair is open.
Gss.9.1 O-family bosons for which (for the ground state) $n_{-2}=0$ or $n_{-1}=0$ transfer charge.
Gss.9.2 At least 1 $o(−2) particle has charge symbolized by Q'=−n/3, Here, 4≥$≥1. Here, n=1, 2, or 4. At least 1 $o(−1) particle has charge symbolized by Q'=+n/3. For each $, the charge of $o(−2) is the negative of the charge of $o(−1). Particles $o(0) have 0 charge.
Gss.9.3 O-family bosons for which (for the ground state) $n_{-4}=0$ or $n_{-3}=0$ transfer property-3.
Gss.9.4 O-family bosons for which (for the ground state) $n_{-6}=0$ or $n_{-5}=0$ transfer magnetic moment.
Gss.9.5 O-family bosons for which (for the ground state) $n_{-8}=0$ or $n_{-7}=0$ transfer property-1.
Gss.9.6 For the w-, h-, and o-families, closed oscillator pairs within the range IOM(−8;−8,3;3) correspond to channels. No other channels pertain.
Gss.9.7 Basic particles for which $\Omega<0$ cannot range freely.
Gss.9.8 O-family basic particles are created in (at least) pairs or triplets.
Gss.10.1 One trio of s-family bosons provides for gluons pertaining to quarks people consider to be matter. The other trio pertains to quarks people consider to be antimatter.
Gss.11.1 For the l-family, combinations of the 4 solutions correspond to 2 of the 3 possible members of an L=1 set (the M=0 member does not apply) and to the 1 member (M=0) of an L=0 set.
Gss.11.2 For the q-family, for S=1/2, combinations of the 4 solutions correspond to 4 of the 5 members of an L=2 set (the M=0 member does not apply).
Gss.11.3 For χ an even positive integer, each of the $| n_\chi , n_{\chi+1} >$ states denoted by $| -1 , -2 >$ or by $| -2 , -1 >$ corresponds to spin/ħ = 1/2.
Gss.11.4 Q-family basic fermions for which S=3/2 have either $n_4=-1$ and $n_5=-2$ or $n_4=-2$ and $n_5=-1$.
Gss.11.5 If an S=3/2 basic fermion absorbs charge, the fermion becomes a quark or a charged lepton.
Gss.11.6 If an S=7/2 basic fermion absorbs charge, the fermion becomes a quark or a charged lepton.
Gss.12.1 For observed astrophysical objects of above some size, era Eó correlates with era Zύ, for 1≤ó=ύ≤3.
Gss.12.2 1e2468& repels astrophysical objects from each other. 2e246& attracts astrophysical objects to each other. 3e24& repels astrophysical objects from each other.
Gss.13.1 For some $ (with 4≥$≥1) and some ύ (with ύ=1, 2, or 4), Q'=−ύ/3 for $o(−2) and Q'=+ύ/3 for $o(−1).
Gss.13.2 People consider interactions that convert anti-quarks into quarks (or vice-versa) to violate P symmetry.
Gss.13.3 O-family bosons ($o(−2) and $o(−1), for some 4≥$≥1) mediate interactions that exhibit P violation.
Gss.13.4 Some o-family members $o% for which 4≥$≥1 and %=−2 or −1 contribute to CP violation approximately the amounts for which the Standard Model can account.
Gss.13.5 Other o-family members contribute to CP violation, beyond that for which the Standard Model can account.

Gss.13.6 Phenomena people associate with axions exist. People can associate such phenomena with e-family coherences.

Gss.13.7 Differences between ύ-symmetry for IOM(−8;−8,9;9) and ύ-symmetry for IOM(−2;−2,3;3) correlate with sizes of ύ-violations people do not associate with Standard Model physics. Here, ύ can be (at least) C, P, CP, or T.

Gss.14.1 In the sense of traditional perturbation models for interactions between elementary particles, people might consider that q-, s-, and o-family particles traverse space-like trajectories between vertices.

Gss.16.1 For charged leptons (either M'=−3 or M'=+3), people can benefit by correlating the range −1≤M"≤3 with an L=2 system.

Gss.16.2 For the L=2 system that includes charged leptons, $m(M'',-3) \propto e^{M''\zeta''}(1+d(M''))$, in which −1≤M"≤3, d(0)=d(2), and d(−1)=d(1)=d(3)=0.

Gss.16.3 The up-to-3 neutrino masses correlate with up to 3 numbers of the form m(M",0), in which ύ−3≤M"≤ύ, for which M"=ύ correlates with the largest mass not ruled out by observations.

Gss.16.4 The formula for m(M", M') has meaning for M"<−1. The trigonometric-like pattern for d(M") continues throughout the range −6≤M"≤3. d(M",0) = 0.

Gss.17.1 For photon-graviton series basic bosons, for interactions between 2 M"=0 leptons, the relative vertex strength per channel follows a pattern established by the relative vertex strengths per channel for photons and gravitons.

Gss.17.2 Vertex strengths scale per particle properties.

Gss.17.3 For interactions between 2 electrons, the strengths of 3e4& and 3e24& are roughly equal at a particle separation of λ_3.

Gss.17.4 For interactions between 2 electrons, the strengths of 2e6& and 2e246& are roughly equal at a particle separation of λ_2.

Gss.17.5 For interactions between 2 electrons, the strengths of 1e8& and 1e2468& are roughly equal at a particle separation of λ_1.

Gss.20.1 People can use the matching of oscillators and symmetry labels to characterize a symmetry.

Gss.20.2 People can consider any ensemble to be (relative to itself) its own ensemble@1.

Gss.20.3 Asymmetry in the spatial distribution of ensembles@24#1 might correlate with the existence of ensembles@24#2.

Gss.21.1 Q-family basic fermions for which $n_2=n_3=-1$ (with respect to interactions with the w-, h-, and o-families) do not interact directly with 4e2& (photons).

Gss.21.2 Q-family basic fermions for which $n_2=n_3=n_4=n_5=n_6=n_7=-1$ (with respect to interactions with the w-, h-, and o-families) do not interact directly with any e-family members other than 1e%& for which 8∈%.

Gss.23.1 People might consider that incompatibilities exist between Ω<0 and each of traditional special relativity and traditional general relativity.

We list suggestions for observational and experimental research

SOR.6.1 Verify (to a smaller than current experimental uncertainty-range) or refute β' = β.

SOR.8.1 Detect instances or effects of, or rule out (to some confidence level), 3e24& coherence between photons and gravitons.

SOR.8.2 Measure or infer signs and magnitudes for forces mediated by e-family members other than 4e2& and 3e4&.

SOR.9.1 Determine the extent to which strengths of interactions mediated by w- , h-, and o-family basic bosons correlate with the concept of channels.

SOR.9.2 Verify or rule out (to some confidence level) existence of o-family bosons.

SOR.9.3 Determine properties (such as charge, mass, and magnetic moment) of o-family bosons.

SOR.9.4 Determine ranges for o-family forces.
SOR.9.5 Verify or rule out (to some confidence level) that o-family bosons cannot be created singly.
SOR.9.6 Determine or rule out (to some confidence level) non-zero binding energies for pairs and triplets of o-family bosons.
SOR.9.7 Determine the extent to which o-family bosons provide for the strong interaction's varying from R^0 spatial dependence.
SOR.9.8 Verify or rule out (to some confidence level) changes to nuclear theory people propose based on o-family physics.
SOR.10.1 Determine the number of channels that pertain for gluon-mediated (or s-family-meditated) interactions.
SOR.11.1 Detect or rule out (to some confidence level) the existence of basic fermions for which S=3/2 or S=7/2.
SOR.11.2 Rule out (to some confidence level) or detect the existence of basic fermions for which S=5/2.
SOR.11.3 Measure or infer properties of S≥3/2 basic fermions.
SOR.11.4 Detect or rule out (to some confidence level) the existence of interactions that convert fermions between S≥3/2 n-type and S=1/2.
SOR.11.5 Verify or rule out (to some confidence level) q- and l-family interaction rules we show regarding the w-, h-, and o-families. Determine strengths for interactions for which strengths are yet to be determined.
SOR.11.6 To what extent does either n-type model (n-type-S or n-type-D) better correlate with observations than does the other n-type model?
SOR.12.1 Estimate strengths and directions (attraction or repulsion) for e-family forces other than 4e2& and 3e4&.
SOR.12.2 Estimate charges of objects for which 3e24& currently dominates.
SOR.13.1 Detect or rule out (to some confidence level) that $o(−2) and $o(−1) bosons can covert a q-family S=1/2 fermion from anti-quark to quark and vice-versa.
SOR.13.2 Rule out (to some confidence level) or detect that $o(−2) and $o(−1) bosons can covert a q-family S=1/2 fermion (quark) to an l-family fermion (lepton) and vice-versa.
SOR.13.3 Determine the extents to which w-, h-, and o-family-mediated interactions (and/or e-family-coherence-mediated interactions) account for observed P violations, CP violations, or other such violations.
SOR.15.1 Verify or rule out (to some confidence level) that much of the difference between the W-boson mass we calculate and the observed W-boson mass correlates with a non-zero magnetic moment for W bosons.
SOR.16.1 Measure neutrino masses.
SOR.16.2 Rule out (to some confidence level) or detect the existence of M"≥−2, M'=0 particles.
SOR.18.1 Determine dependences of neutrino-oscillation rates on influences of mass or gravity.
SOR.19.1 Find or rule out (to some confidence level) evidence (other than currently assumed evidence) of effects on baryonic matter of dark matter.
SOR.20.1 Determine the extents to which fermion properties vary by ensemble.
SOR.20.2 Determine limits on similarities and differences between ensembles@5 matter and ensemble@1 matter.
SOR.20.3 Use data about ensemble@1 to infer w-, h-, and o-family interaction strengths and/or big-bang phenomena related to w-, h-, and o-family-mediated interactions.
SOR.20.4 Determine the extents to which fields related to various 3e%&, 2e%&, and 1e%& bosons have impacted CMB radiation.
SOR.20.5 Determine the extent to which observations (for example, about the uniformity of distribution of ensembles@24#1 stuff) cannot be explained by traditional physics and/or work in this paper directly related to ensembles@24#1. Estimate the extent to which people might

attribute any such unexplained phenomena to the presence of ensembles@24#2 (for example, via effects early in the big bang).

SOR.20.6 Detect or rule out (to some confidence level) effects on ensembles@24#1 of ensembles@24#2.

SOR.20.7 Determine the extent to which ensembles contribute equally to actual densities of the universe. (Here we differentiate actual from inferred {for example, inferred via CMB data}.)

SOR.22.1 Perform observations and experiments sufficient to allow people to decide among at least 4 hypotheses regarding S≥3/2 fermions, ensembles, and non-baryonic matter. [items (22.3), (22.4), (22.5), and (22.6)]

SOR.23.1 Detect or rule out (to some confidence level), say for 3e4& and 2e6&, fields analogous to the magnetic field people associate with 4e2&.

SOR.23.2 To what extent might non-4e2% e-family fields analogous to the magnetic field correlate with jets associated with quasars or with other phenomena?

We list suggestions for theoretical research

STR.4.1 Complete mathematics related to type-1 solutions sufficiently to describe wave functions for edge cases for D_p = 3 and ν = −3/2. (Possibly, extend the work to pertain to edge cases for other D_p and ν.)

STR.4.2 To what extent might people derive benefit from IOM for any 1 or more than 1 of the following? D can be an integer < 1. D can be other than an integer. 2ν can be other than an integer. $\pm_\chi$ can be other than +1 or −1.

STR.4.3 To what extent would it be useful, for D=3, for people to consider a quantum number s such that S=(s−1)/2? (Here, S(S+1)=(1/4)·(s^2−1). Here, perhaps, s is any non-zero integer.)

STR.4.4 Explore IOM for cases in which D_e is even and D_p is even.

STR.4.5 How might people improve or extend the technique for cataloging quantum approaches?

STR.5.1 To what extent would people find beneficial defining and using a value for a D_{*e}?

STR.8.1 To what extent might people benefit by considering that, for IOM(−2;0,9;9), the 0-and-1 oscillator pair correlates with a fold in an otherwise flat energy-momentum space (or space time), that oscillators −2 through 0 correlate with 2 or 3 flat energy-like (or time-like) dimensions, and that oscillators 1 through 9 correlate with 8 or 9 flat momentum-like (or space-like) dimensions?

STR.8.2 To what extent might people find it appropriate to associate $(t')^0$ behavior with $e%& interactions? (Here, t' denotes time.)

STR.8.3 Harmonize models and observations or experiments regarding S for e-family members.

STR.9.1 How best might people explore the existence and characteristics of $o% particles?

STR.9.2 What known or new phenomena people might explain based on the o-family?

STR.9.3 Estimate properties of o-family bosons.

STR.9.4 Estimate ranges for o-family forces.

STR.9.5 To what extent do o-family bosons correspond to aspects of the shell model for atomic nuclei? (Harmonic-oscillator math seems to pertain to each of the o-family and the shell model.)

STR.9.6 To what extent might people explain properties of atomic nuclei, based on o-family bosons (and gluons and other physics)?

STR.9.7 To what extent might people explain properties of neutron stars, based on o-family bosons (and other physics)?

STR.9.8 To what extent might people benefit by exploring the notion that o-family bosons erase and paint fermion properties? (Here, we have in mind possible parallels to people's considering that gluons erase and paint color charge.)

STR.9.9 Harmonize theory and observations or experiments regarding numbers of channels for interactions mediated by w-, h-, and o-family basic bosons.

STR.10.1 Develop enough theory to enable experiments to determine the number of s-family channels.

STR.10.2 To what extent might people benefit by exploring the possibility that the spatial dependence of s-family forces is R^0 (asymptotic freedom) and that o-family (and possibly h-family) bosons provide for variation of the strong force from R^0 spatial dependence?

STR.10.3 To what extent might people find it appropriate to associate $(t')^{-2}$ behavior with 4s% interactions? (Here, t' denotes time.)

STR.10.4 To what extent might people benefit by considering the possibility that, if \$s(−4,−2) and \$s(−3,−1) coherences exist, the spatial character of the related force could be R^2? (Here, we have in mind a possible series - 1e2468& ↔ R^{-8}, ..., 4e2& ↔ R^{-2}, 4s% ↔ R^0, ...)

STR.11.1 Estimate masses and magnetic moments for basic fermions for which S≥3/2.

STR.11.2 Describe possible composite objects (such as nuclei or atoms) for which S≥3/2 basic fermions would be components.

STR.12.1 Predict strengths and directions (attraction or repulsion) for e-family forces other than 4e2& and 3e4&.

STR.12.2 To what extent do neutrinos interact with \$e%&-for-which-2ϵ% bosons based on, for example, neutrinos being transformed into virtual pairs, each consisting of a charged lepton and a 4w2 or 4w3?

STR.13.1 To what extent do \$o%-mediated interactions correlate in strength with measured CP and/or P violations?

STR.13.2 To what extent do some o-family-mediated interactions correlate in strength with CP and/or P violations for which the Standard Model can account?

STR.13.3 To what extent might people benefit by considering that an interaction of a fermion with a 4w1 (or a 4w2, or 4w3) basic boson erases - from the fermion - 1 generation (or charge) and paints - on to the fermion - 1 generation (or charge)?

STR.13.4 To what extent might people benefit by considering that an interaction of a fermion with a \$o(0) (or a \$o(−2) or a \$o(−1)) basic boson erases - from the fermion - 1 generation (or charge) and paints - on to the fermion - 1 generation (or charge)?

STR.13.5 To what extent might people benefit by considering that an interaction of a fermion with a \$o(−4) or a \$o(−3) basic boson erases - from the fermion - 1 unit of property-3 and paints - on to the fermion - 1 unit of property-3?

STR.13.6 To what extent might people benefit by considering that an interaction of a fermion with a \$o(−6) or a \$o(−5) basic boson erases - from the fermion - 1 unit of magnetic moment and paints - on to the fermion - 1 unit of magnetic moment?

STR.13.7 To what extent might people benefit by considering that an interaction of a fermion with a \$o(−8) or a \$o(−7) basic boson erases - from the fermion - 1 unit of property-1 and paints - on to the fermion - 1 unit of property-1?

STR.14.1 To what extent might people benefit by considering that $\Omega<0$-for-D_p aspects of IOM correlate with space-like (or faster-than-light-speed) behavior?

STR.15.1 Estimate a value for the magnetic moment of W bosons.

STR.15.2 Develop a model specifying values of D+2ν correlating with o-family masses.

STR.15.3 To what extent might people benefit from considering that an antiparticle for each of 4w1, 5h1, and the various \$o(0) particles (4≥\$≥1) has an Ω that is the negative of the Ω for the particle?

STR.16.1 What would be the impact of detection of M'=0 baryonic-matter fermions having M"≥−2?

STR.16.2 To what extent do masses of q- and l-family fermions correlate with masses of w- and h-family bosons?

STR.16.3 To what extent does the appearance, in a formula for the ratios of masses of charged leptons, of a power of the ratio of the square roots of 2 lepton masses correlate with the possible applicability of the Koide formula? [(16.58) and (16.61)]

STR.16.4 Develop a model specifying values of D+2ν, ύ(M"), and adjustments to use for calculating quark masses. [items including and following item (16.71)]

STR.16.5 Extend work regarding STR.4.1 to model S=1/2 basic fermion masses.

STR.16.6 Develop models people can use to estimate masses for S≥3/2 basic fermions.

STR.17.1 Develop theory sufficient to predict choices - attraction, repulsion, or neither - for each e-family interaction between 2 particles.

STR.17.2 To what extent might people benefit by exploring theories in which the property of charge correlates with concepts of curvature for a space time in which the number of spatial dimensions is 1?

STR.17.3 To what extent might people benefit by considering that effects of $e%& interactions might be modeled in a flat space time for which there are no more than 3 time-like dimensions and no more than 2(5−$)+1 space-like dimensions? (Here, the term flat refers to a Minkowski-like metric with each off-diagonal term having a value of 0. Here, one can consider 4 cases - $ = 4, 3, 2, and 1.)

STR.17.4 To what extent should people associate phenomena related to $e%& with people's notions of space-time froth? (For example, do 4e2&-related phenomena provide for supposed froth on a scale people associate with the Planck length?)

STR.17.5 To what extent might people use parallels to the electromagnetic vector potential when describing gravity and other $e%& interactions? (For example, for 3e4&, property-3 could be an analog to charge; and, a concept of a property-3 current could lead to a property-3-based analog to magnetic fields.)

STR.17.6 To what extent might people extend the Standard Model to include gravity and non-traditional e-family interactions? (For example, can people base such an extension on potentials, currents, and so forth suggested by STR.17.5?)

STR.17.7 Develop a suitable IOM perturbation theory (possibly based on something like Feynman diagrams) for e-family interactions.

STR.17.8 Use such an IOM perturbation theory [STR.17.7] to estimate magnetic-moment anomalies. [items (6.47) and (6.48)]

STR.19.1 How best should people try to directly detect matter not associated with baryonic matter?

STR.20.1 How might people detect or rule out (to some confidence level) the existence of ensembles@24#2?

STR.20.2 To what extent should theories related to gravitation, curvature of space time, general relativity, and so forth be reexamined because of the existence of 4 similar or 8 total forms of gravity (3e4&)?

STR.20.3 To what extent does trying to account, in the Einstein field equations, for dark energy via the cosmological constant suffice?

STR.20.4 To the extent unchanged Einstein field equations pertain, to what extent does the cosmological-constant term vary (as a function of position in space time) because of clumping of dark-energy stuff?

STR.20.5 For each of the various eras ($Z_{ύ}$, per Section 12), which interactions most affect neutrinos?

STR.20.6 To what extent might people use the term universe to denote ensembles@48 or just ensembles@24#1?

STR.21.1 Estimate ratios of densities of the universe for baryonic matter, dark matter, and dark energy, assuming that S≥3/2 basic fermions provide all dark matter and dark energy.

STR.22.1 Review observational data and calculations of the density of baryonic matter (equivalent to a mass density of approximately 9.9×10^{-30} g/cm^3, which is similar to mass density of 5.9

protons per cubic meter) to find a basis for determining the extent to which S≥3/2 fermions might pertain. [Ref.22.1]

STR.23.1 To what extent would people benefit by correlating IOM(−1;−1,2;2) with Dirac-equation spinors? (Each has 4 components. Here, D_e=D_p=2.)

STR.23.2 To what extent would people benefit by considering Q'=0 to be a charge, such as Q'=−1, and not just a lack of charge?

STR.23.3 To what extents might people consider that traditional-physics special relativity and traditional-physics general relativity do no precisely apply at distances for which o- and s-family interactions have significant influence?

STR.23.4 To what extents might people consider that traditional-physics special relativity and traditional-physics general relativity do no precisely apply (for a non-zero-mass system) at distances smaller than roughly a system's spin/mass length [(6.15)]?

STR.23.5 Estimate probabilities of significant perturbations to the evolution of earth, to planetary motion, or to other solar-system phenomena, because of dark-matter stuff.

STR.23.6 To what extent might people, based on this work, change estimates of the number of independent physical constants?

STR.24.1 Develop more-coherent or more-compact (than this paper shows) models that produce or improve on results this paper offers.

STR.24.2 Develop theory people can use to produce models (such as this paper shows or STR.24.1 suggests).

STR.24.3 To what extent can people benefit by developing hybrid models that include aspects of IOM and the Standard Model?

STR.24.4 To what extent can people develop theory that generates adequately useful, possibly hybrid models (that include aspects of IOM and the Standard Model)?

STR.24.5 For what other applications might people use IOM?

Section 26 References

Ref.4.1 Wolfram Alpha, computational knowledge engine, Wolfram Alpha LLC, http://mathworld.wolfram.com/DeltaFunction.html.

Ref.6.1 Particle Data Group, Electroweak (web page), *The Particle Adventure*, Lawrence Berkeley National Laboratory, http://www.particleadventure.org/electroweak.html.

Ref.6.2 G. T. Adylov, et. al., A measurement of the electromagnetic size of the pion from direct elastic pion scattering data at 50 GeV/c, *Nuclear Physics B*, Volume 128, Issue 3, 3 October 1977, pages 461-505. (http://dx.doi.org/10.1016/0550-3213(77)90056-6)

Ref.6.3 T. Quinn et al, Improved Determination of G Using Two Methods, *Phys. Rev. Lett,* 111, 101102, 2013. (http://link.aps.org/doi/10.1103/PhysRevLett.111.101102)

Ref.6.4 J. Beringer et al. (Particle Data Group), *Phys. Rev. D86*, 010001 (2012). (http://pdg.lbl.gov/2012/reviews/rpp2012-rev-phys-constants.pdf)

Ref.6.5 J. Beringer et al. (Particle Data Group), *Phys. Rev. D86*, 010001 (2012). (http://pdg.lbl.gov/2012/tables/rpp2012-sum-leptons.pdf)

Ref.12.1 N. G. Busca, et. al., Baryon Oscillations in the Lyα forest of BOSS quasars, arXiv:1211.2616 [astro-ph.CO].

Ref.12.2 A. Riess, et. al., Type Ia Supernova Discoveries at z > 1 from the *Hubble Space Telescope*: Evidence for Past Deceleration and Constraints on Dark Energy Evolution, *The Astrophysical Journal*, 607, 665 (2004). (doi:10.1086/383612) (http://iopscience.iop.org/0004-637X/607/2/665)

Ref.14.1 Wolfram Alpha, computational knowledge engine, Wolfram Alpha LLC, http://mathworld.wolfram.com/Laplacian.html.

Ref.15.1 J. Beringer et al. (Particle Data Group), *PR D86*, 010001 (2012) and 2013 partial update for the 2014 edition (URL: http://pdg.lbl.gov). (http://pdg.lbl.gov/2013/tables/rpp2013-sum-gauge-higgs-bosons.pdf)

Ref.15.2 CMS collaboration (2012). "Observation of a new boson at a mass of 125 GeV with the CMS experiment at the LHC". *Physics Letters B* 716 (1): 30–61. arXiv:1207.7235. Bibcode:2012PhLB..716...30C. doi:10.1016/j.physletb.2012.08.021.

Ref.15.3 ATLAS collaboration (2012). "Observation of a New Particle in the Search for the Standard Model Higgs Boson with the ATLAS Detector at the LHC". *Physics Letters B* 716 (1): 1–29. arXiv:1207.7214. Bibcode:2012PhLB..716....1A. doi:10.1016/j.physletb.2012.08.020.

Ref.15.4 J. Beringer et al. (Particle Data Group), *PR D86*, 010001 (2012) and 2013 partial update for the 2014 edition (URL: http://pdg.lbl.gov).

Ref.16.1 J. Beringer et al. (Particle Data Group), *Phys. Rev. D86*, 010001 (2012). (http://pdg.lbl.gov/2012/tables/rpp2012-sum-quarks.pdf)

Ref.16.2 J. Beringer et al. (Particle Data Group), *PR D86*, 010001 (2012) and 2013 partial update for the 2014 edition (URL: http://pdg.lbl.gov). (http://pdg.lbl.gov/2013/tables/rpp2013-sum-leptons.pdf)

Ref.16.3 S. Thomas, F. Abdalla, and O. Lahav, Upper Bound of 0.28 eV on the Neutrino Masses from the Largest Photometric Redshift Survey, *Phys. Rev. Lett. 105*, 031301, 2010. (http://arxiv.org/abs/0911.5291)

Ref.16.4 A. Melchiorri, Constraints on Neutrino Physics from Planck, European Space Agency, http://www.rssd.esa.int/SA/PLANCK/docs/eslab47/Session06_CMB_Cosmology_and_Fundamental_Physics/47ESLAB_April_04_17_30_Melchiorri.pdf.

Ref.18.1 J. Beringer et al. (Particle Data Group), *Phys. Rev. D86*, 010001 (2012). "13. Neutrino mass, mixing, and oscillations," page 46. (http://pdg.lbl.gov/2012/reviews/rpp2012-rev-neutrino-mixing.pdf)

Ref.19.1 J. Beringer et al. (Particle Data Group), *Phys. Rev. D86*, 010001 (2012). (http://pdg.lbl.gov/2012/reviews/rpp2012-rev-dark-matter.pdf)

Ref.19.2 J. Beringer et al. (Particle Data Group), *Phys. Rev. D86*, 010001 (2012). (http://pdg.lbl.gov/2012/reviews/rpp2012-rev-cosmic-microwave-background.pdf)

Ref.19.3 Mark Peplow, Planck telescope peers into primordial universe, *Nature News*, Nature Publishing Group, March 21, 2013. (http://www.nature.com/news/planck-telescope-peers-into-primordial-universe-1.12658)

Ref.22.1 Wilkinson Microwave Anisotropy Probe, http://wmap.gsfc.nasa.gov/universe/WMAP_Universe.pdf

www.ingramcontent.com/pod-product-compliance
Lightning Source LLC
Chambersburg PA
CBHW080305180526
45167CB00006B/2672
* 9 7 8 1 4 9 4 8 4 9 8 2 5 *